P9-CEU-461

How to Be
a Better Lover...
Longer ℘

OTHER BOOKS BY DEBBIE DRAKE

Dancercize

*Book of Exercise,
Diet and Beauty*

*Debbie Drake's
Secrets of Perfect
Figure Development*

How to Be a Better Lover... Longer *by* Debbie Drake

Prentice-Hall, Inc.
Englewood Cliffs, N.J.

How to Be a Better Lover . . . Longer
By Debbie Drake

Copyright © 1973 by Debbie Drake
Illustrations copyright © 1970 by Diversified Products Corp.

All rights reserved. No part of this book may be
reproduced in any form or by any means, except
for the inclusion of brief quotations in a review,
without permission in writing from the publisher.

Printed in the United States of America

Prentice-Hall International, Inc., London
Prentice-Hall of Australia, Pty. Ltd., North Sydney
Prentice-Hall of Canada, Ltd., Toronto
Prentice-Hall of India Private Ltd., New Delhi
Prentice-Hall of Japan, Inc., Tokyo

Library of Congress Cataloging in Publication Data
Drake, Debbie.
    How to be a better lover . . . longer.
    1. Sex instruction for men.   I. Title.
HQ36.D7   301.41'76'32   72-10363
ISBN 0-13-402172-X

To FRANCIS BAKEWELL, S.J.
WHO HAS GIVEN ME A NEW
AND DEEPER UNDERSTANDING
OF THE BEAUTY OF SEXUAL LOVE
and who has made his lifework the bringing of this
understanding to couples who seek his help

ACKNOWLEDGMENTS

This book was a long time in the making, and it would not have been possible without the cooperation of many, many people.

First my thanks to writer Ann Morell, who blushed a lot but was able to translate my thoughts and feelings into words; to Hank Reduta, Manager, the Chicago Tribune–New York News Syndicate, who blushed not one bit, but gave me some fine insights and viewpoints; and to Hollis Alpert, whose suggestions were invaluable in the planning stages of the book.

I am deeply grateful for the interest shown by the following men and for their help in gathering data on the physical, psychological, and sociological aspects of sex: Dr. Edward C. Smith, psychiatrist; Dr. Lewis Barbato, psychiatrist; Dr. Paul Knott, psychologist; Dr. Fred Todd, psychologist; Dr. H. M. Muffly, gynecologist; Dr. A. Myron Lawson, dentist; Dr. R. Chris Weatherley-White, plastic surgeon; Dr. Lloyd Lunston, internist; Dr. Eugene Beyer, dermatologist; Dr. James A. Peterson, psychologist; Dr. Jim James; Dr. Kenneth H. Cooper; Dr. Paul Hunsinger; and Dr. Jim Kauffman, Dean of Student Life, University of Denver.

For their contributions to new directions in men's fashion,

I want to thank Sid Mayer, Merchandise Manager, Saks Fifth Avenue; Neil Fox, Vice-President and Merchandise Manager, Neiman-Marcus; Bernie Schwarts, owner, Eric Ross; Frank Reilly, President, Wallach's, Inc.; Neusteters of Denver; Eddie Stevens of Miami, and designer John Weitz. For his help in men's hair care and styling, my thanks to Roger de Anfrasio, of Roger of New York.

Very special thanks go to Prentice-Hall—to Nicholas J. D'Incecco for his encouragement and help, and to Dennis Fawcett, my editor, whose patience and understanding meant so very much to me.

I also want to thank my agent, Dick Rubin, and Frank Page, Bob Siedleman, Dorset White, Marty Connelly, Brett Marshall, Marty Lipton, Ed Turley, and Jim Storey for their contributions and cooperation.

To Mike Douglas, Merv Griffin, Johnny Carson, Bob Hope, James Garner, Andy Williams, James Stewart, Tom Poston, Steve Lawrence, Steve Allen, Morey Amsterdam, Dick Clark, Glenn Corbett, John Wayne, Jim Mahoney, Warren Cowan, Kaye Stevens, Tina Louise, Phyllis Diller, Carla Braelli, Linda Crystal, Dorothy Manners, and Helen Gurley Brown, my deep thanks for their gracious giving of their time and thoughts.

My thanks are also due all those men and women from all over the world whose questions about living and loving made me sense the need for this book. And, finally, to the girls who fly with TWA, Continental, Frontier, and United, a big thank-you for sharing their many flying hours and thousands of opinions about men.

INTRODUCTION

Sex is the sport of kings and plumbers, of queens and nurses, of dukes, duchesses, and dishwashers. It is the favorite sport of bees, of birds, of "goldfish in the privacy of bowls."

Sex, the songwriters would have us believe, is "doin' what comes naturally"—but is it?

Language barriers have prevented me from interviewing the female of the species among birds, bees, and goldfish—but I've certainly talked to thousands of women on the subject, and these interviews have prompted me to write this book.

Women want you, need you, love you, but the thousands of women I've talked with say that they would select only about 5 percent of you males to be on ALL-AMERICAN teams in the great sport of sex. Few of you, it seems, know the types of passes that make touchdowns with women.

All of you men know exactly how you want members of *our* team to look, act, walk, or run, and you want us to know all the rules that stir *your* chemistry. Why is it that so few of you know what a *woman really* wants in you?

Can you honestly say that you know what turns a woman on—and keeps her turned on so that she falls in love and *stays* in love with you? To name a few things, a woman wants a man to have good health, reasonable good looks, charm, and the know-how to make a woman feel like a princess. Women want men to know how to wear clothes well, to be vibrant and interesting. But most of all, women want men to be thoughtful lovers, and at least adequate sex partners.

Now the songwriters must be wrong because, if I can believe what thousands of women tell me, all this isn't "doin' what comes naturally."

Since there are no schools where such things are taught, I have written this book to help you men master these fine arts, to evaluate your current assets, capitalize on your good qualities, eliminate your weaknesses, and make you into the sort of men women really want as lovers.

Men who receive only compliments from women as they play the game of love may never know the truth about themselves.

Ego feeds on flattery, but real growth comes from following constructive criticism and trusting someone who will give it to you.

I know the position in which most women find themselves. They try to build you up because they like you, so they often give you half-truths. As much as I like you men and enjoy telling you about your nice qualities, I always feel that I'm leaving the job only half done.

I want to say, "You are nice, and you have many wonderful traits, but I would like to help you achieve all the charm, sophistication, and poise that is possible for you."

I know that every man can be more charming and more handsome.

Even a ninety-pound weakling can develop the physique of a Greek god—if he's willing to work at it. Puny men can build themselves up, and overweight men can certainly trim themselves down.

So many physical inadequacies, too, can be corrected, sometimes medically, sometimes by plastic surgery.

My aim is to make you as attractive as possible, and to make you a more thoughtful and adequate lover. Wouldn't you like to be the kind of man to whom women want to return for seconds—for thirds—for a lifetime of loving?

I am saddened when I meet a man who feels that money, position, and prestige alone can buy physical, psychological, and chemical attraction. They can't. But knowledge and determination, and the benefits of good training, can.

Even though I've enjoyed the company of many sophisticated, charming, witty, and intelligent men, I could count on two fingers the number of men whom I've considered as marriage material.

If I were the only woman who felt this way, I wouldn't speak so positively, but over the years I've spoken to thousands of women who have expressed these same sentiments.

The most frustrating part about the situation is that I know that most of the problems that ruined the attractiveness of these men could have been corrected. I strongly feel that today, for the first time, a great many men *are* open-minded and are searching for ways to become more attractive. Perhaps you men have come to the realization that you have paid more attention to your golf or tennis game than you have ever paid to turning on your wife or lover—and keeping her turned on!

In the following pages I'm going to share with you a great many of the ways in which you can improve so that you can make home runs with women, instead of being a no-hitter.

With some good coaching, almost every man can make it. Remember, those who play the game of sex badly usually get kicked off the team. The divorce rate in this country reflects a scorecard of players who struck out in bed.

I might add that regardless of how you feel about "women's lib," sociologists are already pointing out that the movement is having far-reaching effects on "Love, American Style." The same woman who once grabbed the first thing in pants to save her from the horrible fate of being an unmarried female in our society is now demanding a lot more. These women are more casual about their affairs, less eager to marry a man who doesn't measure up—and, in short, they are going to hold out for the near perfection that men have long demanded from women!

But back to so-called married bliss.

The problems are all documented if you want to argue. A highly esteemed group of researchers in Detroit asked women who had been married for fifteen or twenty years how well satisfied they were with their husbands and their marriages.

The Detroit investigators found that only seven out of every hundred women were completely happy with what went on in their marriage after fifteen years. That leaves 93 percent dissatisfied!

There are other studies—by the dozens.

One group from Ohio University studied American men and found that most of their marriages were *devitalized* after fifteen or twenty years of marriage and that there was no sexual ecstasy, much less old-fashioned devotion, left for these couples. They had a sexual life all right—but not with each other!

The truth of the matter is that females, far from being the weaker sex, have a greater physical and emotional capacity for lovemaking than males. They can reach multiple orgasms before a man reaches a single climax. Scientific evidence indicates that a woman's orgasm is generally of greater intensity than a male's, and lasts longer. And, if you're still not impressed, many women admit that they achieve their most intense orgasms through masturbation!

Goodness knows how many women have never yet experienced a single orgasm while having "normal intercourse." How sad! And how unnecessary!

Imagine if dear old Frank knew that except for his take-home paycheck each week, he could be replaced by a $3.98 vibrator—well, almost!

Now, I'm not saying that men are responsible for all the errors committed in the bedchamber, but there is a truism that says a good violinist can get a pretty fair melody out of even a substandard instrument. Yet if you put a Stradivarius into the hands of an amateur, you'll still get a scratchy, uneven performance.

I know that some women are unresponsive and puritanical in their sexual attitudes—and little wonder. They have been conditioned over the centuries to be undemanding, timid, and passive in bed lest they be thought to be loose, over-sexed, or, even worse, lascivious. I personally don't think there's a woman among us who wouldn't respond, and passionately, to a man trained in "major league" love techniques.

You may want to know how I've qualified myself to train men who can't make it in the minor leagues to be a Number One draft choice in the major leagues.

That's a legitimate question. You must have confidence in me if I'm going to improve your batting average in the ball game. In preparation for writing this book, I've studied scientific and literary references, from Kinsey to Shakespeare to Johnson.

I've read *J* and *M* and even considered K and X. And Z and Y. I've interviewed men and women from the smallest towns in the corn belt to the largest metropolitan areas on both coasts.

I've talked with gynecologists, urologists, psychiatrists, psychologists, and marriage counselors, and I've done my homework with the so-called sex experts.

Included in my interviews for this book were leading fashion experts from New York and Europe, dermatologists from Eastern and mid-America, hair stylists who cater to the jet set and to Hollywood—and to the average guy in Kansas City. I have discussed eyewear with optometrists, clothes with the nation's leading tailors, voice with drama coaches, and body movement with physical culture experts.

But most important, I've talked with the women who know you best. I've interviewed them in beauty salons, during public appearances across the country, and in marriage counseling centers. The overwhelming and consistent picture that emerged from my surveys is that women want you to be better lovers.

My surveys have prompted me to write this book. I'm going to teach you HOW TO BE A BETTER LOVER . . . LONGER!

Contents

	Introduction	viii
1	*Husbands—And Other Lousy Lovers*	1
2	*How to Rate Yourself as a Lover—Let It All Hang Out!*	7
3	*Sexability*	17
4	*Work Out to Make Out*	59
5	*Improve the Package—Improve Your Chances*	101
6	*Sex—The Sport Every Man Can Excel In*	153
7	*The Superstars of the Sex League*	169
8	*What's Happening in Your Bedroom?*	181
9	*The Sex Machine*	193
10	*How to Wine and Dine a Date More Effectively!*	197
11	*How to Tell if She Really Loves You*	209
12	*How to Get In—And How to Stay In*	225
13	*How to End a Romance—Without Getting Shot*	235

CONTENTS

14   *What Kills Love?*   249

15   *How to Be a Perfect Husband—
Although No One's Perfect*   265

16   *What Is This Sex Thing All About?*   279

Chapter 1
Husbands–And Other Lousy Lovers

As much as I hate to tell you this, millions of women went to bed last night feeling terribly let down, terribly unfulfilled, and all because you, their partners in the wonderful sport of sex, weren't well enough prepared to make them happy.

Men are the most important element in women's lives. And this, I think, is why women have been so reluctant to tell you the truth, for fear of bruising your egos. Our mothers may have told us to "play the game" and never to appear less than delighted with your lovemaking. But this Mama's girl has seen enough unhappiness and downright misery to call a halt to the ruse.

I feel it's time to reveal what so many women have confided in me, so I'm going to be completely honest with you.

If you're completely honest with yourself and are convinced that you're among that 5 percent who would make the all-American team when it comes to satisfying a woman, you can lend this book to your best friend and return to the game of the week on television.

But you there—not so fast!

I just had a heart-to-heart talk with that cute little blonde you say you made out so smashingly with last night.

1

Her report on the action was that it was more smashed than smashing, since you consumed two martinis before dinner, wine with dinner, and a few highballs after dinner —"just to get in the mood," as you put it!

Her version was that as a lover, you stank, and I'm afraid she meant that literally as well as figuratively. No athlete would consider drinking before playing, and a good lover should certainly take his sport just as seriously. If you fumble, fumble, fumble in the bed, you won't make a touchdown with any woman! And the sad part about the whole situation is that so many men like you are often so tranquilized by drink that they're not even aware of their fourth-string performance!

If you can take it, I'll tell you precisely what your "conquest" of last night related of your performance.

But wait, you say you must have been great—because you felt her dig her fingernails into your back, you heard her moan in ecstasy and sigh in rapture?

Well hooray for her! She may still be stuck in the typing pool, but at least her two years in Actor's Studio weren't a complete washout!

She told me that the fumes she inhaled from your boozing "just to get in the mood" would have made her flunk the friendly state highway patrolman's breatholator test. She mentioned in passing (after almost passing out from the mere memory of your distillery aroma) that your after-dinner cigar was so vile that she almost trounced out of that intimate little restaurant alone—before the two of you had even made it as far as your apartment. But being a natural optimist, she went along with the game.

Then, really warming up to her subject (no, not you), she sighed a *real* sigh, not one of her parlor-game ones, and let go.

"Most men are such lousy lovers, Debbie. Honestly, if you lined up all the fumbling incompetent, inconsiderate men end to end, they would stretch from Manhattan to . . . to . . . to . . . oh, hell—I think I'd just as soon let them lie there end to end—as long as the ends involved are theirs, and not mine!

"Debbie, why can't men learn that you stroke a cat to make it purr, and that a woman needs to be coaxed, stroked, petted, and made love to with undivided attention before she ever can be really turned on? And that once a woman is turned on, she's not easily turned off?"

To be perfectly honest, your partner of last night did have one small compliment to pay about your performance after you plopped your drink down beside the TV and pounced on the Posturepedic with her.

"Just think, Debbie," she recalled, "I didn't have to miss a single second of *African Queen* because his entire action took place during the Twenty Mule Team Borax commercial!"

Now, darling, do you think I'm hitting below the belt to tell you all this?

Just don't attempt to schedule a rematch. Your partner of last night said she'd rather while away her evening with old movies, crewel embroidery, or a yoga class at the "Y."

In fact, her parting shot, if you can take it, was that the lotus position is much more provocative than any of your fumbles of last night!

I know you don't make friends and influence people by being brutally honest all the time, but there are times when complete honesty is absolutely essential, and this is one of them. That's why I must say *most men are lousy lovers*. In fact, I know from talking to so many unhappy women that what you *don't* know about lovemaking could fill a book.

I've talked to them all, wives and sweethearts, trim, tanned matrons and hopeful fiancées, disillusioned suburbanites and swinging career girls.

Their complaints are legion:

"Would he ever call if he didn't want to jump in the sack?"

"He expects me to look like a page out of *Vogue* but I

3

swear, he'd escort me to a presidential inauguration looking like a page out of *Field and Stream!*"

"If I hear one more story about superboy at the office,
I'll scream! I *do* happen to have a pretty interesting job,
too—but do I ever get equal time to talk about *my* work?"

"Why does he have to talk of his past romances?"

"He's incapable of making a conquest—with anyone."

"He's practically a ninety-pound weakling."

"Ha! I worried about getting pregnant—but he's got a
pot that makes *him* look six months gone!"

"I know he bathes, but his feet—whew!"

"Oh, he's turned on to women all right. Somewhere he
read that to caress women is to turn them on. He's good
for at least thirty minutes of cuddling, then sixty seconds
after insertion, he's made his home run—and once again,
I've gotten all warmed up for nothing!"

"The only time we ever seem to make love is after he
picks a nasty fight about something I've done. Then I'm
supposed to be so grateful for his 'making up.'"

"He ruined a good thing by blabbing to all his friends. I
mean, I'm certainly no Victorian, but don't I deserve a little better than that?"

"What's his rush? I like three-minute egg timers for char-
ades or Scrabble, but for sex, I'd rather use an hourglass!"

"He's so easily persuaded that I've scored—he never really
asks."

"If he gave a damn about how I feel, he wouldn't scoot
out of here right after we make love. What's he got—a late
date?" I make divine champagne breakfasts—if he'd ever
stick around until the next morning so he could find out!"

And so on . . . And so on . . . And so on. . . .

But winners aren't made by negative thinking.

I know I can make you a better lover, a superlover. No
matter what your current status, I know I can teach you
many things that will make women want to stay with you
for a lifetime of loving.

I'd be very, very happy if I could feel that one by one, I
could change the thousands upon thousands of men about

4

whom women complain into exciting, vibrant partners in the sport of sex.

I'd like to feel I could continue talking to women about their health, their figures, and their *happy* sex life—and never again hear that nationwide lament "Men Are Lousy Lovers."

Chapter 2
How To Rate Yourself As A Lover—Let It All Hang Out!

Our first step in helping you improve as a lover must be a very frank inventory of all your current assets and liabilities.

Are you willing to put yourself in my eager hands? Let's go into the privacy of your bedroom. No wives, roommates, telephones—just intimacy.

The first thing I'm going to ask you to do is to take all your clothes off. What better way is there to get to know each other in the shortest possible time? We can't have any secrets from each other if we're going to succeed in this training camp for all-American lovers.

THE LOVE INSTRUMENT

Since your body is the instrument with which you make love, let's take stock of what you have to work with . . . or play with. I'll sit over here in the corner while you hie yourself over to that full-length mirror. (I do hope we can maintain our privacy, because someone who popped in on us might not understand what we were doing.)

Mmmmmm . . . you're all stripped now and standing

7

in front of that mirror. Women do this all the time, you know. We look ourselves up and down after our baths, and we make sure that we are not putting on an excess inch here or there—why? So we can be more attractive to the most important people in our lives . . . you!

POINTS OF VIEW

What do you see in your mirror?

I do hope you'll be just as honest with yourself as we are when we take our vital statistics. And, please, remember that all my comments are strictly from a woman's point of view . . . or perhaps I should say from *so many women's points of view.*

Let's start at the top and take a good look at your hair. Men have repeatedly told me that nothing turns them on like shining superclean hair on a woman. You men all breathed a collective sigh of relief when we quit all the spray-net mess and went for a more natural, fresh look.

What about your hair? I'd like to run my fingers through it right now and feel that it was shampoo-fresh, with no greasy kid stuff, nothing keeping my hands from caressing the real you!

Even if your hair is thinning, it's a fact of life that frequent shampooing makes for a fuller-appearing head of hair, while oil and most lotions give your hair a less luxuriant look.

Balding a bit? Don't cringe—those male hormones of yours are just asserting themselves! But since we're being so honest with each other, can't you see that your receding hairline is taking a lot of your appeal away? And that every bald spot detracts from your sex appeal? You aren't being overly vain if you investigate hair transplants, hair weaving, and hairpieces! (I've already done your homework for you, so just check in Chapter 5 to learn how easy and how painless it is to take years off your looks.)

Everyone but you sees you in three dimensions, not just

8

as you see yourself in the mirror while shaving. Grab a hand mirror, now, and survey your hair style from all angles. That's it—now you can see yourself the same way most people do—from the sides or from the back. You never know when an attractive young gal might be following you, so look at your hairstyle the way she does.

By the time we get to Chapter 5, you'll have learned a few tricks with your hair so that it looks great from all angles.

Are you beginnning to feel more comfortable about our togetherness session?

Remember, I'm just saying out loud all the things that most women (your wife, your lover, your secretary, that shapely little neighbor of yours) *think* about you.

ANOTHER ANGLE

Let's look at your ears now. You think they're not important except for hearing? Look again! I once knew a prominent lawyer who had ears that stuck so far out from his head that he should have sued his own mother for not having them surgically corrected when he was still a preschooler! I don't think I could even tell you the color of this fellow's eyes because my attention was riveted on those pitcher ears. His ears may not have hindered him in the courtroom, but they really make a case against his love life? Ears that stick out can easily be corrected by plastic surgery in a short, relatively painless operation, even on adults.

THE EYES HAVE IT

While every single feature of your face is important, the eyes certainly have it over every other feature—or should! Women melt at those baby blue orbs of Paul Newman, dream that Richard Burton's eyes followed them

9

across a crowded room and—yes—would obey the commands flashed by those arrogant, yet soulful, eyes of Yul Brynner. (But you still can't make me take back what I said about bald pates.)

Where does a Wally Cox come in? I'll tell you—right behind his spectacles, that's where. Oh, glasses sometimes arouse a mother image. If your lenses are thick enough, perhaps little old ladies will offer to help *you* across busy streets. But for the most part, glasses are more likely to bring out the mother than the tigress instinct in women.

Your eyes are the mirror of your soul, and if you're obscuring the mirror with the bifocal corrections, you should try better solutions!

If you haven't tried contact lenses—please do.

If you tried contacts years ago and weren't able to adjust to them, try again. There is a whole new world in contacts, from the new "soft" lens to new tints, to shapes designed for special kinds of eye problems, such as astigmatism. And aside from the cosmetic benefits of wearing contacts, there are very practical reasons for contact lenses. Did you know, for instance, that athletes have turned to contacts because unlike ordinary glasses, contacts give greater side vision? Or that surgeons and skiers are now wearing contacts because they don't "fog up" like regular glasses? Or that the wearing of contacts actually is beneficial to some optical irregularities, even arresting progressive farsightedness?

Do give contacts a try. You're excused only if you're in that minority of 5 to 10 percent of persons needing corrections who have been told by an ophthalmologist that contact lenses are not suited to their needs.

If you are stuck with specs, at least avoid the "Mr. Peepers" syndrome. Wear dashing new shapes like ski star Jean Claude Killy, or like the "Today" show's Frank McGee. Or try the supermogul effect with really bold, masculine frames. Whichever the choice, make your glasses an accent, not an apology!

EXCESS BAGGAGE?

Now that I've had my say about your eyes, I might mention that some men really ought to keep their glasses on . . . that is, if the frames are hiding a lot of saggage and baggage around the eyes. Folds of tissue above and below the eyes (and women get them, too) can make you look like an owl who overindulged.

One of my favorite television cohorts had this problem. His eyes (or rather, the flesh above and below his eyes) made him look far older than his forty-five years.

One day Bruce returned to the studio from a week-long holiday in Miami. He appeared to have shed years from his age. His secret, I later learned, was a nearly painless hour-long session with an eminent plastic surgeon who skillfully removed Bruce's excess baggage. I'm willing to bet Bruce sharpened his career-and-caress rating by a good 50 percent, and he's still reaping the benefits of his inexpensive plastic surgery!

PROMINENT PROBOSCIS

Assuming your eyes now have it, can we move on to the nose?

Horses may win by a nose, but men are more likely to lose by a nose. You might tell that "schnozzy" friend of yours that his nose will become more prominent as he grows older. I'll let you in on another secret. That very pretty young thing you've been traipsing around with lately had her nose job done last January, and her pretty little nose now functions a lot better for minor needs like breathing and smelling, too. Oh, she felt sensitive about telling people, until she thought of explaining that the carving job was done to "cure the sinuses"—and it did!

11

LIGHT UP AND SMILE

Toothpaste ads may be overdone, but honestly, is there anything more attractive than a nice smile? Of course you want your mouth kissing sweet, your teeth showing absolutely no sign of coffee or cigarette stains. And if you must wear precious metals, wear them anywhere but in your mouth! No matter what that dental work cost, let's keep it as unobtrusive as possible.

Now, take your index finger and rub it gently across your lips. I hope you find your lips soft, relaxed, and kissable. That's the way I'd like to find them!

Before we leave that handsome face of yours and move on down toward the playground, pick up a hand mirror and get a good look at your chin. You're not stuck with a weak chin, you know—or a lantern-jaw effect—any more than you are stuck with oversized ears, bags under the eyes, or any of the other problems we've discussed. I'm certainly not getting a rebate from any plastic surgeons, but what "Dr. Michael Angelo," whose real name is available from your local medical society, can do to correct your facial faults would amaze you!

BASIC BASES

I'll just touch a few of the important bases on the way to the playground. All right?

BASE ONE: I hate to say this, but please remember that I'm doing this all for your own good. You know, when I first saw you, I thought you looked . . . well, a little predatory. That chin-forward stride may look great as you burst into the conference room, but baby, it doesn't do a thing for any woman. Reserve that aggressive stance for sales meetings, and relax a bit when you're trying to score. Women are pushovers for lithe, limber, good posture. We like to feel a man is strong, but we like a supple, easy grace, one that's rather slow, yet controlled.

BASE TWO: We're at your chest, and I hope it's so nice that I would like to rest here for a while. I share most women's feelings about men whose chests are so scrawny that they look as though they've just been through a death march. Chest hair? I personally don't care a whit if you are shaggy as an Afghan hound or totally hairless. By the time we get as far as your chest, I couldn't care less, as long as that chest is solid, well filled out, and strong.

BASE THREE: "And he took her in his arms . . ."

Wait! Women want to be protected. Your arms have to be at *least* bigger than hers or she won't feel one teensy bit protected. You don't have to have twenty-one-inch biceps, but I would suggest at least a good sixteen-inch biceps—if you want to have the upper hand (and arm) in your lovemaking. Keeping a few ten-to-forty-pound dumbbells at arm's length isn't such a bad idea, either, as you gain in years. Arms once muscular from lots of exercise can become a bit flabby unless some form of exercise is maintained. You don't want to look like one of those pot-bellied, flabby-armed old has-beens we see along the Riviera, do you? (Unless you've got a few million in a Swiss bank to help make up for the deficit!)

LOW-SCORE ERROR

I do hope you're not guilty of the worst fault of all—the rubber foam tummy. The girth that goes great for department store Santas probably (outside of *bad breath* and *smelly feet*) turns off more women than any other defect. No kidding! You fellows want *our* tissues soft but firm and in the right places; what makes you think we don't feel the same way about yours? A soft, protruding pot is a real no-no. And if that doesn't scare you, this should!

Carrying around that extra blubber means you're putting a strain on your heart—and expending a lot of energy which you might blissfully have used in bed or for other fun activities in years to come. Although I hate to tell you this, you won't live as long as your slender friends, and if you don't *live* as long you won't *love* as long, and we certainly don't want *that* to happen!

I shouldn't have to tell you that women aren't enchanted by very skinny lovers either. Keeping trim is one thing, but without muscle you are not big-league material, friend!

THE BATTER'S BOX

Below your stomach are your tools of love, your penis and the family jewels. If you're like most red-blooded American boys, you grew up surreptitiously sneaking looks at other boys in the gym locker room and worrying about the size of your penis.

What a waste of your time! So many men seem to feel that their penis is smaller than it should be. They don't realize that small penises tend to enlarge much more during the act of love than do organs that are naturally a little larger in an unexcited state!

One very candid woman told me, "Debbie, a man's penis is big enough if it reaches the body of the woman he loves. It's what he does with what he's got, not the size of the thing, that really counts.

"As for men who are self-conscious because they feel their penis is too small, why, most women are so busy worrying about what the man thinks of the size and fullness of their breasts that they wouldn't notice the size of their lover's organ unless they tripped over it!"

What matters as far as lovemaking is concerned is how well your penis is educated, whether you can move it after insertion (not just thrusts—that's elemental, but vertically and horizontally so that it caresses and tugs the tissues of the vagina and stimulates the clitoris!).

A well-muscled young lover I interviewed told me that many of his friends still worry about the size of their playing equipment. He gave these worriers two hints. "Tell them never to look down at their penis via their stomach," he said. "Other people don't look at him like this, so why should he? Instead have him look into a full-length mirror so he can get a more accurate picture of how others see him."

He added, "A man can make what he has look longer by cutting back his pubic hair. Often the hair is so thick and curly that trimming it down seems to add a half-inch or more to the size of the organ."

You learn something every day! (But puh-lease be careful about wielding scissors in that precious, precious area!)

Now, while you're still in the nude, it is wise to do a skin check, all over. Gently scratch your upper arms, your thighs. Does the flaking of dry skin make the air look like a snowstorm in a paperweight? Dry, scaly skin is awfully unattractive. It looks bad physically, and it feels bad to the woman you love. If you have a dry skin problem, get your druggist to suggest soap and lotions that help. A fragrant bath oil after your shower will correct the dry skin and will make you smell nice all over, too.

BEFORE YOU WALK OFF

Let's linger for just a moment on those appendages with which you play footsie. You don't think feet are exactly a sex symbol? This might surprise you then: One of the most common complaints that women have about shortcomings in their sex partners is, bluntly, that a lot of you he-men have feet that are less than clean! Worse, say the gals, a lot of you have feet that smell as though they'd been running the 440 for a week in—oh, shame—the same pair of socks.

Those feet are for more than walking, and be sure you keep them immaculately clean, and free of calluses and corns.

Who, after all, wants to play footsie with a horned toad?

15

I think we've covered all the bases now, from your head to your toes. Can you honestly say you shape up to my ideas of what a lover should look like? Of course, I'm just one woman, but the things I've said reflect what so many women have told me they want in the man they love.

You can unlock the bedroom door now because I have completed my look at you. Those few faults we found can certainly be corrected, and starting on the next page, you'll find the know-how to put a new and more handsome image in that mirror of yours!

Chapter 3
Sexability

If we were still in the Garden of Eden (before Eve succumbed to the wicked suggestions of that nasty serpent) and were running around as God created us, just how would you shape up?

How many Eves would come running after you if you were stripped nude, without your jazzy new car, your two-hundred-dollar suit, or all those lovely plastic cards that buy expensive evenings out on the town? How big a fig leaf (or how many of them) would it take to cover up your flaws? Sexability is not having to use a fig leaf.

Would Eve rate your sexability high—from her very first look at your manly shape—or would she rather play with her garden than play with any of your attractions?

UNDER YOUR FIG LEAF

Your sexability is the sum total of your masculinity, your virility, your potency, your stamina and vigor, and the ability of your body to keep up with the most erotic passions of your mind.

17

UNDER HER FIG LEAF

What attracts you to a woman? Aren't her physical attractions the first thing you notice? Don't you rate a woman tops in sexability when she displays a gorgeous body, curves in the proper places, a full bosom, a slim waist, and agile and beautiful hips?

Your body turns her on the same way, you know. If you have good proportions that proclaim, "I'm a healthy, virile male," she is going to be attracted to you. Your body shows how you feel about *yourself*, that you've cared enough to maintain physical fitness, build stamina and vigor, and keep within your daily routine all the elements of good health which give you the promise of providing the best loving a woman ever had!

Your sexability is, honestly, the messages your body sends out. If your body is wonderfully developed, with broad shoulders, a flat tummy, and strong, well-proportioned arms and legs, it's a sign of a high sexability rating. Your body is sending out strong waves of masculinity, signals that reach her before you even begin your first conversation. With a great body, you're almost scoring before your mind begins to *think* about scoring!

WATCH THE PROS

You can build up your sexability, increasing it greatly, in just a season of working with me.

I will demand only forty-five minutes of your time, three times each week.

Just consider how often you've sat in a football stadium and watched the players move their beautifully coordinated bodies up and down the field. Then consider the seasons and seasons of practice, the thousands of hours of kicking and tackling, of conditioning and wind sprints, of body building and proper dieting that these superathletes have invested in their total performance.

If you will spend a tiny fraction of your week with me, I can guarantee that you will realize many, many benefits from our program.

You may not end up starring in a football game, but in a game that's a lot more fun! All I can promise is that if you follow my sexability program you will—

FIRST, build a body that gives you a masculinity and vitality for longer and more passionate lovemaking;

SECOND, give yourself a health insurance policy that pays off in allowing you to make love for as long as you want to;

THIRD, give yourself a life insurance policy that will delay the aging process and insure you a longer and better life, since I know we can keep your *physiological* age younger than your *chronological* age;

FOURTH, increase your mental vigor and, at the same time, increase your physical abilities so that they are at least equal to your wonderful brain power!

FIFTH, learn how to become and stay a healthy individual in an environment that is becoming increasingly antagonistic toward your health.

If you will just donate a small fraction of the time you spend sitting watching other athletes perform, or lighting and smoking cigarettes, or consuming an evening beer, or standing over a barbecue pit watching high-cholesterol pork ribs cook, we can increase your masculinity, your physical fitness, and your SEXABILITY!

I know we can do this in a season, even if you're one of those people who "don't have the time" to work at a physical fitness program to improve your sexability.

ARE YOU PHFFFFT?

I see so many men who don't have time to work at being physically fit that they just go physically "phfffft" in-

19

stead; their wives and lovers are the ones who suffer terribly as a result.

Think of all the seasons it takes to build a marvelous athlete, and of all the time a very good athlete puts into his training program.

Now think of all the seasons you've already wasted, with no body and sexability to show for it!

SEXABILITY HIJACKERS

I hate to have to tell you this, but unless you're in an isolated rural area, you're going to have to work harder to preserve your physical health and prevent premature age than people did just a few generations ago. You've got ? lot more forces stacked up against you.

You have air pollution from industry, from your beloved automobile, and from cigarettes (which emit up to thirteen thousand times more carbon monoxide than the industrial air we breathe). You have noise pollution . . . cars and planes and air-conditioners and rock music. (Your dentist tried to help by getting an almost noiseless drill—but then he went and installed canned music all over his office!)

You have pep pills to get you started in the morning, and tranquilizers to calm you down at the end of the day. You have hundreds of brands of cigarettes and liquor to choose from when you want to dissipate. You even have a hangover pill to cure the effects of too much alcoholic indulgence. You have dozens of brands of throat lozenges to help your smoker's throat, and a city of almost any size has a cancer treatment center for those who won't kick the cigarette habit.

You've got preservatives that rob your foods of their natural goodness and lots of insecticides to spray into the air you breathe.

To support your indulgence in all these "good things"

20

of our way of life, you have a high-paying, high-pressure job—and the high blood pressure to go with it.

If you think all this contributes to preserving your youth and good looks, I'm afraid all the bad elements of our way of life have already caused you some brain damage! Simply to protect yourself from the pollutants and the pace of the planet world, circa the 1970s, you've got to work much harder than your grandfather ever did.

A LOOK AT YOUR LIFE-STYLE

True   False

1 I am no more than ten pounds overweight. (See chart, Chapter 4.)

2 I have stopped cigarette smoking. (If the black-bordered notice on the packs didn't convince you, you're in bad shape.)

3 I do two different types of body-building exercises every day.

4 I understand good nutrition, balanced diets, and the dangers of overpartaking of certain types of food.

5 I drink no more than one cup of coffee daily.

6. I do not consume an alcoholic drink every day.

7 I know what to do for tension without resorting to tranquilizers or sleeping aids.

8 I have had a complete physical examination, including ECG, in the last year (in the last six months for those over thirty-five). If I have any heart problems, my doctor has prescribed a diet-and-exercise regimen, and I follow it to the letter.

9 I walk upstairs whenever I can, and leave the elevators free for the old and infirm.

10 I walk whenever I can, and I'd never think

21

of getting the car out just to drive three blocks to the store.

Scoring Your General Fitness

100% *True* answers:	Congratulations! You're a candidate for the all-American team in the great game of sex.
80%–90% *True* answers:	You've got just a few changes to make in the next thirty days of our working out to make out. If you had to answer "False" to even one of these questions, you're not as healthy as you could be. If you said no to question 8, make an appointment for a physical examination right this minute so that you can start our program as soon as possible.
60%–70% *True* answers:	Your life-style is making you a borderline case for both living and loving. Promise yourself to reform immediately, and you'll see a big change in thirty days.
Less than **50%** *True* answers:	I've found you just in time! You may have to work harder than some other men in my program, but you'll also see a bigger improvement as we work together to change the way you live.

HEAP BIG SMOKE—BUT NO FIRE

Smoking can burn up your sexability in two ways.

First, you can literally smoke yourself *out of* bed.

Second, you can literally smoke yourself *into* bed—one of those permanent beds that they lower into the ground as your final resting-place.

You don't want to smoke yourself *in* or *out* of bed if you're the man I think you are. I'll go over both unpleasant alternatives (either one of which destroys your sexability) just in case you're not convinced.

SMOKING YOURSELF OUT OF BED

As far as smoking yourself out of bed, there are several ways you can do this. There is the simple fact that a smoker's breath is as appealing as Smog City, U.S.A. on an overcast day. If you don't believe me, I just wish you could abstain from your tobacco habit for a few weeks (to get back your sense of smell and taste) and then crawl into bed with a woman who just crushed out her forty-eighth cigarette of the day. No amount of brushing, mouthwashing, or gum chewing can clear the breath of a heavy smoker. The smoke aroma lingers on . . . although the smoker seldom does.

You can also smoke yourself out of bed, because with each cigarette you cut down on your lung capacity and elasticity. You build up steam as you approach orgasm, you know, and you're going to need all the lung power you can muster if you plan to be in the love game for some time to come. (Of course, there's always the possibility that you'll never get her into bed in the first place because of your unsexy crow's-feet around your eyes. Smoking ages you in more visible areas than your lungs, you know. Heavy smoking does to a man what a decade of excessive sunning does—it wrinkles the skin around the eyes so that you look like a reject from a prune factory. She won't be able to tell about your general health, nor things like lungs and stamina, until you get on very intimate terms. But if you look ten years older than your real age, how attractive are you physically?)

Those men in the cigarette commercials may look sexy, but just how sexy does a woman find a man who *really* smokes heavily? I've known quite a few heavy smokers, and here's how their habit affected me.

THE HACKER

Hacker would cough through dinner, although he continued to light up between courses. He coughed during movies, and he often coughed so loudly that he annoyed the people sitting around us. (He was always considerate enough to notice this, however, and would leave to stroll out to the lobby and have a cigarette.) He coughed in some decidedly more intimate situations, too, and kept them from being a lot more intimate!

THE WHEEZER

I was flattered the first time he came to pick me up, because I thought he must have run all the way from his apartment to mine! When his wheezing and shortness of breath didn't subside, however, I realized that it was his constant smoking that caused the wheezing and that the wheeze was part of his permanent packaging.

THE SADIST-MASOCHIST

There are millions of sadist-masochist smokers in the United States. My friend, prominent in television production in the Northeast, brought home to me what a terrible addiction smoking is and how it hurts not only the smoker but those around him as well.

Years of smoking resulted in cancer of his vocal chords, which had to be removed surgically when radiation treatment failed. The strain on his young wife and family was terrible—not to mention the strain on the victim himself, who survived the ten-hour operation but emerged voiceless.

With larynx, pharynx, and a portion of his food tube removed, he breathed through a surgical hole at the base of his throat. Speech was impossible for some months, so he communicated with a pad and a pencil.

During this period, as I visited him, I was almost sickened by the sights in the hospital. Several of his roommates had to "feed" themselves by pouring a formula in a glass tube which they held high over their heads so it could pass through thinner tubes which led through their noses to their stomachs. Several had entire jaws, mouths, or tongues missing. I went home after each visit with an intense depression, a depression made even worse because I knew smoking had been responsible for almost every case on my friend's hospital floor.

To say my friend was a sadist-masochist may sound strong.

But he had been warned by his doctor, fully five years before the onset of his cancer, that his heavy smoking would damage either his heart or his lungs. He continued to smoke.

His wife begged him to quit "for the children's sake." He continued to smoke.

25

A team of doctors performed the surgery to save his life. He learned, via intensive therapy, to force his diaphragm to push air so he could form a sort of speech through an artificial voice box. He was able to resume his career. Yet, dropping in at his office one day, I was amazed—and sickened—to surprise him as he lit a cigarette and pressed it to the breathing hole at the base of his neck in an attempt to "inhale" the cigarette! After all the harm he had done himself and his family, and after all the medical efforts made in his behalf, he continued to smoke.

I can't have any feelings of love for a man who's a slave to any habit—unless that habit is me.

All smokers are a little selfish, since they make litter in people's living rooms as they pile ashtrays high with cigarette butts, offend others with their smoker's breath and smoke-smelling clothes, and pollute the very air others breathe.

At the very least, the smoker is advertising to others that he's careless about his finances—that he literally has money to burn!

I wish you'd toss every cigarette, lighter, cigar, and pipe in the nearest Trash-Masher right this minute and never, never ever think of smoking again. Then you could read the following warnings with a much easier conscience:

> SMOKERS who are heavily addicted to their habit are twice as likely to die between the ages of twenty-five and sixty-five.
> SMOKERS run 1.7 times the risk of developing coronary artery disease as do nonsmokers.
> SMOKERS who reach middle age and continue to smoke heavily run a 50 to 200 percent greater risk of coronary artery disease.
> SMOKERS are far more likely to develop lung cancer, since research shows that 80 to 90 percent of all lung cancers are directly traceable to smoking.
> SMOKERS who escape lung cancer can always wait

for killing emphysema to develop in their smoke-damaged lungs.

SMOKERS who escape both lung cancer and emphysema might get off with chronic bronchitis caused by the decreased size of the air passages due to smoking.

SMOKERS usually suffer from chronic fatigue, because the cigarettes they inhale cause sudden elevation of the blood pressure and blood sugar level. A quick boost, and a quick letdown. After two packs of boosts and letdowns daily, the smoker's poor system is exhausted.

SMOKERS (ah, aren't you glad you tossed all that smoking equipment away?) are likely to be proved less potent than nonsmokers. As this research comes to make headlines, won't you be glad you quit?

As far as those sexy cigarette ads we see in our slick magazines, I'd like to see a few with models from the cancer hospital. Can you imagine a couple, drifting hand in hand through a sunlight forest, holding cigarettes, looking adoringly at each other—he at her battery-powered voice box and she at the food tube inserted in his partially present nose? Isn't *that* a sexy way to present what cigarettes do to enhance your love life?

Or how about those "magic moment" cigarette ads? Their party's over, and they're sitting in their cute little apartment, amid the used coffee cups and dishes, smiling at each other over a final evening cigarette. Behind him are two large oxygen tanks and a face mask, treatment for his cigarette-produced emphysema. She's a lovely young thing—until you realize that the designer scarf she wears is held at chin level to hide the devastating effects of throat and esophageal surgery she had to remove the malignant tumors caused by her smoking.

Sexy?

Appealing?

If the ads had to show, in glossy full color, just what smoking does to destroy sex appeal and health, I think

27

every adult would do what the kids are doing about smoking—they'd join the "unhooked generation."

BOOZE IN HER BOUDOIR

I'd love to photograph and write a series of liquor ads that are just as "sexy" as the ones we're exposed to in our magazines.

Bad Scene One

The first full-color ad would be of a young couple. She's on her pretty feet, sipping at a glass of amber-hued liquid. He's reclining on the couch. The furnishings are beautiful. She is beautiful. He's handsome—but a little puffy around the eyes and a little red in the face (or is that sunburn?). He's holding out a glass. He's saying, "WHILE YOU'RE UP, WILL YOU GET ME A LIVER TRANSPLANT?"

Bad Scene Two

This one is a group of affluent cocktail party people, commonly called the Beautiful People, gathered in a posh New York penthouse setting. They all look terribly chic and each partygoer is holding to his or her lips a glass of colorless or bright orange or bright red liquid. The headline reads: "IT LEAVES YOU BRAINLESS!"

Bad Scene Three

This could be set in several locations. The first is the emergency room of a hospital. Internes are working over the inert form of a man who's been badly mutilated in a one-car contest. Actually, three factors were involved: his overimbibing at a party; his late-model car, loaded with the speed that kills; and an innocent telephone pole situated at a point on a steep curve.

The headline: "I'LL JUST HAVE ONE MORE FOR THAT TEN-MINUTE DRIVE HOME."

There are many more bad scenes, of course. There are

custodial institutions for drinkers who weren't ever able to stop . . . institutions that drain us to the tune of over $25 million every year.

My ad campaign could go on and on, and as the scenes would get worse, so would the characters in them. They might not ever look like skid-row bums. Many of the people who abuse alcohol look very much like people you know, like the man who shares your office (the one with the shaking hands and the early-morning irritability) or your neighbor (who has good skills, but seems to keep losing jobs) or that friend of a friend (whose wife puts up with his drinking as the family and finances go down the drain).

LIQUOR—THE LONG-TERM SENTENCE

We all know you can have your drinks at lunch and your drinks at dinner and never get "drunk" in the skid-row sense of the word.

But I just wish you could measure, over a ten-year period, your vitality and appearance against that of a contemporary who abstained from alcohol! You'll look (with your "moderate" drinking) a good decade older than your friend of the same age who uses his lunch period exercising instead of imbibing. Handling your liquor every night may keep you from getting in trouble with the law—but it certainly won't keep you from aging as much as if you'd been fighting the law every long hard night that you consume alcohol.

If you choose liquor over more healthful recreations, you just won't be the man for the job—or for the girl!

THE "SOCIAL LUBRICANT"

Alcohol in careful quantities has long been toasted as a "social lubricant," but those tiny ounces can add up to make you a perfect washout in your social life, your life-work, and, of course, in BED.

I'd rather see you abstain from alcohol altogether, but if you won't, I'll settle for your promise never to take a drink every day, or even every other day. That way, you're using alcohol as something aside from your daily routine and are much less likely ever to allow it to cause you to wash out as a *lover* and as a *liver*, or to wash out your life and your liver.

Sexperts agree that many a case of impotence began when the male, flushed with alcohol but thinking he was flushed with passion, instigated what he thought was going to be one of the world's greatest love matches. What happened? He struck out so badly in bed that he awakened the next morning with a horrible sense of failure and resolved to make his bad scene good the next day. (At lunchtime he downed a couple of martinis to fortify his resolve and had his predinner cocktails as well. Meanwhile, his poor system was asking, "What's happening?")

That night his lovemaking again fizzled out, and fear set in. A few nights later he was afraid to make love, afraid of reproducing the same failure that was brought on by his first alcoholic binge!

Of course, this probably won't happen to you for a while—I hope.

But why not decide that alcohol isn't going to be part of your daily life before you even run such a risk?

Try these problems on for size—do you need any of them?

> DRINKERS (and this is the very least that can happen) are going to gain a lot more weight than nondrinkers, even if they follow the same diet. Drinkers consume *FIVE TIMES* as many calories in each ounce of alcohol as they would in an ounce of milk. Dry wines and whiskey, as you will note on our diet charts later in this chapter, contain no carbohydrate grams, but they certainly contain the calories. Can you afford the extra calories?

DRINKERS run the risk of undernourishment if they start skipping food in favor of liquor. Alcoholics do great damage to their bodies because they rely on the deceiving calories which exist in alcohol. They miss out on every vitamin and mineral and rob their systems of needed nutrients.

DRINKERS run the risk of developing very high blood cholesterol, since alcohol speedily converts to saturated fat. Do you need to run this risk when you're already fighting so many other toxic substances?

DRINKERS are much more prone to kidney disease.

DRINKERS are much more prone to "large cell" anemia.

DRINKERS are much more prone to liver disease Studies show that with all the other toxic substances in the environment which our body must now fight, cirrhosis of the liver is on the increase, even among the so-called social drinkers. Isn't this reason enough to stop? If not, read on.

DRINKERS, after consuming merely a few drinks, show in tests that their hearing, their vision, and their muscular coordination is less efficient.

DRINKERS and their drinking are directly related to at least *one-fourth* of all automobile accidents. (In at least one large city, this figure was shown to be closer to 50 percent.)

DRINKERS who continue their habit to excess may suffer mental deterioration and disorientation (most likely, experts say, due to the effects of their imperfect nutrition as well as their excessive drinking).

DRINKERS are less efficient in their lifework. They miss an average of over twenty days per year due to illness related to their drinking.

DRINKERS (if we can't appeal to you on a health basis alone) spend almost $12 *billion* a year on alcohol. Considering that no nutritional benefits are derived from alcohol and that steady use of alcohol can lead to impotence and so many other horrible disorders, does this make any sense at all? It's a shame that any of us can still associate "Scotch thrift" with drinking.

THE DISTAFF SIDE ON DRINKING

You might well lend your ears to what women have to say about "lovers" who drink before making love.

Said one: "If he feels he has to be bombed before being amorous, it makes me feel as though I'm an unpleasant task for which he must bolster himself with booze before he can think of making love to me."

Said another: "Many men feel they're so great when they've had a drink too many. I think their passion may be up, but that's the only thing about them that is. Usually it's a losing game when they even attempt love after liquor."

Said a disillusioned wife: "It got so our lovemaking always took place only on Saturday night after a long cocktail party. He thought it was the 'mellow time,' as he called it. It took a marriage counselor and a session with AA before we realized he was trying to make up with drinking what he lacked physically. After he got straightened out, so did our sex. It's sober or nothing, as far as I'm concerned now."

I'd like to make the liquor ads just as sexy as my cancer-hospital cigarette ads. We'd get the models from a local drying-out hospital. We'd catch the red-rimmed eyes and the puffy features and the shaking hands. We'd show how the "social lubricant" can wash love and life right out of a good-looking young man and reduce him to a human wreck, a has-been in the masculinity market.

FOODS FOR POWER PLAY

I'm not going to give you any fad diet, but I am going to show you how you can follow a diet that will make you look younger, feel younger, have more oomph and vitality, and have more drive to pursue the good things in life (like lovemaking!).

We're going to discover the power-packed foods that will beef you up. We're going to cut out all those deceitful calories that give you a quick lift and a fast letdown while adding plenty of extra inches around your waist.

My diet plan will let you gain strength while you lose weight.

But it isn't going to be one of those diets that offer you a banana split for breakfast or two martinis for lunch or a chocolate milk shake for dinner.

The so-called diet planners whose brainstorms you read about in almost every promotion-geared magazine are earning fat profits on the indulgence and laziness of the public. Every overweight person with fifty cents in his pocket picks up a copy of a magazine offering such goodies as "The Amazing Ice Cream Diet." Beware of any diet that offers you a gustatorial joyride!

Such diets leave you temporarily glutted with foods that offer deceitful calories, rob your system of the needed proteins, and make you very, very hungry before the next meal.

I'm giving you many diet alternatives in this chapter.

I'm giving you a LIFETIME VITALITY AND EN-ERGY DIET FOR MASCULINITY which will give you all the vigor and pep you need for total living and loving.

I'm also going to show you how you can "take your diet out to dinner" at an Italian restaurant and keep your body full of the nutrients it needs for vital living, without putting on the extra pounds.

You CAN eat like a king and keep your weight down and your energy up. But PLEASE stop starving yourself for a few days and then going on a carbohydrate binge.

It's very easy to decrease your carbohydrate intake while INCREASING YOUR ENERGY INTAKE. My simple plan teaches you to select foods that make you a stronger man, a man more fit to meet the needs of everyday life.

OH, PROMISE ME . . .

As we begin, please promise me one thing: Let your dieting be a secret between you and your refrigerator. Dieting is a big bore to hear about. Wait until you've taken off the pounds; that's the time to crow about the diet!

I don't want you to be one of those poor souls who are in a constant swivet, wondering, "Let's see . . . if I have one-third cup of green beans, cooked with one pat of margarine, would that add up to more calories than if I had a half-head of lettuce with a teaspoonful of French dressing over it?" You're too busy to occupy your mind with such trivia!

Instead, we'll learn to add up our protein grams so that we consume from 90 to 100 grams of protein in each day's meals. We'll learn to add up our carbohydrate grams so that we consume fewer than 60 carbohydrate grams in each day's meals.

To give you an idea of how this works, just look at my list of some of the DIET HIJACKERS which, when they gang up in great numbers, will blitz your diet plans. You might make a list of these and tape the list to your refrigerator door. There are too many diet hijackers on our supermarket shelves to list them all, but if you're being honest with yourself you can guess that if a slice of chocolate cake is a hijacker, a fudge brownie is likely to be one, too. Get the picture?

The diet hijackers in our meals have several things in common. They're almost all high priced. They're high priced in calories. They're high priced in carbohydrates. They're high priced on your grocer's shelves (ounce for ounce, you're paying several times as much for potato chips as you are for steak, and think of the difference in nutritional value!). They're also high priced in what they cost you in terms of unhealthful fat, more frequent doctor bills, and higher insurance rates.

Let's take a look at some of the DIET HIJACKERS.

Breads and Starches	*Carbohydrate Grams*
Rye, white, raisin, or cracked wheat bread, 1 slice	12
Gingerbread, one 2-inch square	28
Chocolate cake, fudge icing, ⅙th slice of 10-inch diameter cake	70
Corn meal muffin	22
Cake doughnut	17
Macaroni (and cheese), 1 cup	44
Apple pie, one 4-inch slice	53
Mince pie, one 4-inch slice	62
Rice, 1 cup, cooked	44
Spaghetti with meat sauce, 1 cup	36
Candies and Sweets	
Milk chocolate, 1 ounce	16
Fudge without nuts, one 1-inch square	23
Pancake syrup, 1 tablespoon	15
Sugar, 1 lump	7
Gelatin, 1 cup	36
Beverages	
Beer, one 12-ounce can	16
Colas, 12 ounces	42
Ginger ale, 12 ounces	31
Sweet cider, 1 cup	34
Cocoa, 1 cup	26

As we've said before, you are limiting yourself to 60 carbohydrate grams per day. Can't you see how the hijackers will keep you from losing weight? Why, just take one slice of the chocolate cake (with its 70 carbohydrate grams), and you've overshot your daily budget by 10 grams!

TAKING YOUR DIET OUT
TO DINNER

For a further look at what the diet hijackers can do, suppose you went to an Italian restaurant and ordered a pasta. Have you ever seen Italian pasta arrive on anything less than an enormous platter? Here's how your dinner adds up. We'll have to assume you're getting *at least* two cups of a spaghetti-type dish. That's a minimum of 72 carbohydrate grams right there! Suppose you have two slices of rich Italian bread. That's another 24 carbohydrate grams. You eat a large salad with Italian dressing (only 5 carbohydrate grams). You consume a can of beer with your meal (another 16 carbohydate grams).

Your whopping total of carbohydrate grams *for just one meal* is 117 carbohydrate grams—nearly double your quota for the entire day!

You could have gone to that same Italian restaurant and enjoyed the cuisine without the carbohydrates. Now that you're smarter about what carbohydrates do to your diet, here's what you order:

Menu Item	*Carbohydrate Grams*
Antipasto plate (including black and green olives, sliced cucumber, carrot strips)	5
Veal Scaloppine à la Genovese (3-inch squares of veal sautéed in olive oil and simmered with sauterne; seasoned with sage, garlic, and nutmeg; garnished with mushrooms and capers)	4
Medium-size salad with dietetic Italian dressing	2
Dry white wine, 1 glass	0
Herb tea, 1 cup	0
Total	11 carbohydrate grams

Congratulations! You've dined out, had a healthful meal, and consumed less than one-sixth of your budgeted 60 carbohydrate grams per day. You planned so well that you could allow yourself the reward of the glass of wine which, since it's dry rather than sweet, adds no carbohydrate grams to your repast.

People who eat out have no excuse for not being able to follow my diet.

Would your most important client even guess you were dieting if you ordered the following at a restaurant?

Menu Item	Carbohydrate Grams
Scotch and soda (if you must)	0
Salad of lettuce, black olives, julienne ham, turkey, and cheese	4
Thousand Island dressing	1
Jellied consommé	0
Sirloin steak, 6 ounces	0
Buttered asparagus spears	3
Sanka	0
Grand (for your diet) Total	8 carbohydrate grams

You couldn't feel like a martyr after a meal like this, could you? You've held to your diet, yet you've packed away a healthful number of those important protein grams, too. Here's how your protein grams added up:

	Protein Grams
Sirloin, 6-ounce strip	40
Ham strips	6
Chicken strips	10
Swiss cheese strips	5
Total protein grams	61

Because you've consumed a goodly portion of protein, you'll find yourself with plenty of energy, go-power, and vitality. You'll look and feel better, and the healthful foods you're eating will soon result in a healthy-looking body.

You may have noticed that both my suggested dinner meals use a very low percentage of your total allowance of 60 carbohydrate grams. There's a good reason for this. If you consume the greater part of your carbohydrate grams at breakfast and lunch, you're more likely to metabolize them during your day's activities—before they can turn to fat.

Of course, you're still going to have your sugars and starches. You're going to have orange juice (at 26 carbohydrate grams a cup). You'll have buckwheat pancakes (at 6 grams each) some pleasant Sunday morning. You'll have noodles as a starch some lunchtime (at a great big 35 carbohydrate grams per cup). But you'll simply learn to estimate the grams in each so you'll be within your 60-gram allowance every day.

YOUR HEALTH HELPERS

Now I'm going to give you a list of the Health Helpers, the foods and drinks you'll want in building your healthful diet without adding many carbohydrate grams to your allowance. These are the ones you want on your team as you build a diet regimen that you'll find easy to live with, year in and year out.

I don't have space to give you every single food item's count of carbohydrate and protein grams. There are simply too many foods to list. If you really get fascinated with the subject, and feel you *must* know how pickled pigs' feet are going to fit into your daily planning, turn to Uncle Sam. For just $1.50 you can order "Composition of Foods" from the Superintendent of Documents, U.S. Government Printing Office, Washington, D.C. 20402. In its 190 pages, you'll find listed the makeup of every food, from abalone to zwieback, that you'll ever consume. (Pickled pigs' feet, incidentally, contain no carbohydrate grams at all, whether you eat one or ten. They have 75.8 protein grams *per pound,* but almost as many fat grams *per pound,* which make them not the sort of food you're going to consider a staple in your diet.)

The book literally catalogs foods from A to Z and will prove very useful to you in your planning.

YOUR DIET CAN EARN
YOU DOLLARS

Rich man, poor man, beggar man, thief. . . .

You and your diet and your physical condition may add up to the rung you reach on the executive ladder.

A recent report by a national magazine noted that only 10 percent of the men in the $25,000 to $50,000 per year salary category were more than ten pounds overweight.

On the other hand, in the $10,000 to $20,000 per year category, about 35 percent (over one-third) were considered definitely overweight.

Business wants leaders who look lean and trim and ready for wheeling and dealing. The image of a fat executive just doesn't fit into today's thinking.

Of course, you don't want to look scrawny or underfed, either. If your problem is one of overweight or underweight that needs immediate attention and fast work, here are two diets to help you see fast results.*

HIGH-POWERED REDUCING DIET

You can pick and choose your way to weight loss and still consume those important ninety to one hundred daily protein grams on this diet. Read the rules before beginning.

---

* Always check with your doctor first before starting on any diet.

RULES FOR REDUCING DIET

1 Eat lean meats and trim off all visible fat. Meat is preferably broiled or roasted.
2 One green leafy vegetable should be eaten daily, and a yellow vegetable twice weekly. Eat fresh fruits, without sugar added.
3 You may use salt, pepper, vinegar, tea, Sanka, bouillon, or consommé in moderate amounts.
4 Avoid sugars, syrups, candies, white bread, packaged cereals, cakes, cookies, doughnuts, soft drinks, pies, sweetened gelatin desserts, and salad dressings.
5 Quantities of milk recommended should be taken every day.

Breakfast
  Fruit: high vitamin C content, 1 serving (Group A)
  Egg: 1; or 1 small serving of other protein foods (Group B)
  Bread: whole grain, 1 slice, and 1 tsp. butter or margarine
  Sanka

Luncheon
  Soup: low calorie, 1 cup (Group C)
  Meat: lean, or other protein food, 3-½ oz. (Group D)
  Vegetable: low calorie, cooked, 1 serving (Group E)
  Salad: no salad oil or sugar (Group F)
  Milk: skim, or buttermilk, 8 fl. oz.
  Sanka

Dinner
  Soup: low calorie, 1 cup (Group C)
  Meat: lean or other protein food, 3-½ oz. (Group D)
  Vegetable: low calorie, cooked, 1 serving (Group E)
  Salad: no salad oil or sugar (Group F)
  Bread: whole grain or enriched, 1 slice; or 1 medium potato
  Butter or Margarine: 1 level tsp.

Milk: skim, or buttermilk, 8 fl. oz.
Fruit: low calorie, 1 serving (Group A)
Sanka

Evening-snack   (9 to 10 P.M.)
Milk: skim, or buttermilk, 8 fl. oz.

FOOD GROUPS

GROUP A, FRUITS
   (UNSWEETENED)

(Raw, water packed; canned, no sugar added; frozen, sugar added)

Grapefruit juice	½ cup
Orange	1 small
Orange juice	½ cup
Strawberries	1 cup
Apple, small	1, 2″ diameter
Applesauce	½ cup
Apricots, canned	6 halves, ½ tbsp. juice
Banana	½ small
Blackberries, raw or canned	½ cup
Blueberries, raw or canned	½ cup
Blueberries, frozen, no sugar added	½ cup
Cantaloupe or muskmelon	¼ of 6″ melon
Cherries, canned	½ cup
Grapefruit, raw	½ small, or ½ cup sections
Grapefruit juice, canned, unsweetened	½ cup
Honeydew melon	⅛ of 2″ melon
Peach, raw	1 medium large
Peaches, canned	4 halves, 3 tbsp. juice
Pear, raw	1 medium
Pears, canned	2 halves, 3 tbsp. juice
Pineapple, raw	⅔ cup
Pineapple, canned	2 small slices, 2 tbsp. juice
Plums, raw	2 medium

Plums, canned	2 medium, 2 tbsp. juice
Prunes, cooked, no sugar	2 medium, 2 tbsp. liquid
Pumpkin, canned	½ cup
Raspberries, raw or canned	½ cup
Watermelon, cubes or balls	1 cup

GROUP B, HIGH PROTEIN FOODS FOR BREAKFAST

Egg	1
Cheese, cottage	½ cup
Cheese, Cheddar	1 oz.
Cheese, Swiss	1 oz.
Cheese, Edam	1 oz.
Bacon, lean, crisp fried, drained	3 strips

GROUP C, SOUPS (LOW CALORIE) (CANNED, DILUTED WITH WATER)

Beef	1 cup
Beef noodle	1 cup
Beef with vegetables	1 cup
Bouillon or consommé	1 cup
Chicken gumbo	1 cup
Chicken noodle	1 cup
Chicken rice	1 cup
Clam chowder	1 cup
Vegetable	1 cup
Vegetable noodle	1 cup

GROUP D, MEAT AND OTHER PROTEIN FOODS

Beef, roast, canned	3½ oz.
Beef, shank (soup meat)	3½ oz.
Lamb, leg or shoulder, roasted	3½ oz.
Veal, leg or shoulder, roasted	3½ oz.
POULTRY	
Chicken, canned, no bones	3½ oz.
Chicken, stewed or roasted, no bones	3½ oz.
Turkey, roasted, no bones	3½ oz.
FISH	
Clams	3½ oz.

Cod steak, baked	3½ oz.
Flounder or sole, baked	3½ oz.
Haddock, cooked	3½ oz.
Halibut steak, cooked	3½ oz.
Lobster, mackerel, salmon, or shrimps, canned	3½ oz.

GROUP E, VEGETABLES (LOW CALORIE COOKED SERVINGS)

Asparagus	⅔ cup cut pieces
Asparagus, green, canned	6 stalks, 2 tbsp. liquid
Beans, snap, green	1 cup
Beets	⅔ cup, diced
Broccoli *	⅔ cup
Brussels sprouts	⅔ cup
Cabbage	½ cup
Carrots *	½ cup, diced
Cauliflower	½ cup
Chard, leaves only *	½ cup
Collards	½ cup
Kale	1 cup
Kohlrabi	½ cup
Onions	½ cup
Parsnips	½ cup
Peas, green	½ cup
Potato	½ medium
Pumpkin, canned	½ cup
Rutabagas	½ cup, diced
Sauerkraut, canned	⅔ cup
Spinach *	½ cup
Squash, summer	½ cup, diced
Squash, winter *	½ cup, mashed
Tomatoes, canned	½ cup
Tomato juice, canned	½ cup
Turnips, white	⅔ cup, diced

GROUP F, VEGETABLES (RAW FOR SALADS)

Cabbage	½ cup, shredded
Carrots	½ cup, shredded

* These vegetables have high vitamin A value. Include at least one serving in the diet each day.

43

Celery	3 small stalks
Chicory or endive	13 to 20 leaves
Cress, garden	5 to 8 sprigs
Cress, water	10 sprigs
Cucumber	½ medium
Escarole	2 large leaves
Kale *	1 cup
Lettuce	5 small leaves
Pepper, green *	1 shell, empty
Pickles, cucumber	1 large dill
Pimentos, canned	1 medium
Radish, red	3 small
Romaine	1 large leaf
Tomato	1 small

WEIGHT-GAINING DIETS
PLAN ONE

Breakfast
2   eggs, bacon or ham or sausage
2   slices whole wheat bread
2   glasses milk
3   tsp. Super Protein

Lunch
2   sandwiches (meat or cheese) on whole wheat bread
2   glasses milk
2   fresh fruit

Dinner
½   lb. steak
2   slices whole wheat bread
2   vegetables
2   glasses milk
3   tsp. Super Protein

Late Snack
3   tbsp. cottage cheese
1   peanut butter sandwich on whole wheat bread
2   glasses milk

Approximate Calories—4,500

* These vegetables have high vitamin A value. Include at least one serving in the diet each day.

44

ADVANCED WEIGHT-GAINING DIET

TO PLAN ONE ADD THE FOLLOWING MEALS

Midmorning
- 2 slices cheese, whole wheat bread
- 2 glasses milk

Midafternoon
- 2 slices whole wheat bread
- 2 glasses milk

Approximate Calories—5,600

First 10 Days

Orange juice—1 cup	110
2 eggs	150
½ lb. steak	650
4 cups milk	660
¼ lb. cottage cheese	100
3 slices whole wheat bread	165
2 patties butter	200
1 serving fruit	75
1 potato	105
Vegetables, cooked	
1 vegetable, raw	
Approximate Total	2,315

Second 10 Days

Orange juice—1 cup	110
2 eggs	150
½ lb. steak	650
4 cups milk	660
¼ lb. cottage cheese	100
4 slices whole wheat bread	220
1 pattie butter	100
2 tbsp. peanut butter	180
1 potato	105
1 vegetable, cooked	100
Fruit dessert	125
Choice of any high protein food	
Approximate Total	2,605

Third 10 Days

¾ lb. steak	975
2 eggs	150
4 cups milk	660
¼ lb. cottage cheese	100
2 tbsp. peanut butter	180
4 slices bread	220
1 pattie butter	100
1 glass orange juice	110
1 vegetable, cooked	100
1 potato	105
Fruit dessert	100
Choice of any high protein food	
Approximate Total	3,230

Last 10 Days

¾ lb. steak	975
2 eggs	150
4 cups milk	660
¼ lb. cottage cheese	100
2 tbsp. peanut butter	180
4 slices bread	220
1 pattie butter	100
1 glass orange juice	110
1 vegetable, cooked	100
1 potato	105
Fruit dessert	100
2 slices cheese	230
Choice of any high protein food	
Approximate Total	3,230

TAKING OFF TEN POUNDS
IN TEN DAYS

1 Breakfast: ½ grapefruit or unsweetened grapefruit juice; as many eggs and bacon slices as you wish; Sanka, herb tea, no cream or sugar.

2	Lunch:	½ grapefruit; meat any style, any amount; salad, as much as you want, with dressing that contains no sugar. Sanka, no cream or sugar.
3	Dinner:	½ grapefruit; meat any style, any amount. You may substitute fish or poultry for meat. Any green, yellow, or red vegetables, as desired. Sanka, no cream or sugar.
4	Bedtime snack:	Tomato juice or skimmed milk.

Important

1 At each meal you must eat until you are full, until you can eat no more.
2 Don't eliminate any menu item. The combination of food is important.
3 Drink only Sanka or herb tea.
4 Do not eat between meals; if you eat the combination of food suggested, you will not need between-meal snacks.
5 Note that this diet completely eliminates sugar and starches. Fat does not form fat, it helps burn it up. You can fry your eggs in butter and use butter generously on your vegetables.
6 Note that with this magic menu you can eat what your family eats except desserts, bread, and white vegetables. While your family eats these, you can have extra helping of meat, salad, and vegetables. You may notice no change for the first few days, then a drop of four or more pounds by the middle of the ten-day period. Alternate this diet with a ten-day diet that includes no more than sixty carbohydrate grams until you achieve your desired weight.

With such a variety of diets, you should be able to select one for your immediate needs (such as quickly peeling off a surplus ten pounds) and one for your future needs (such as keeping your weight at its perfect level while keeping your system full of masculine vitality).

I'll end our talk on diet with one warning.

No woman likes any sign of weakness in a man.

No other bad habit shows so immediately as that of overindulgence in food or lack of exercise. Your well-built body advertises how well you've learned to nourish and exercise the physical equipment you have.

Your body (if it's bulging at the belt and flabby at her focal points) simply says, "THE PERSON WHO LIVES INSIDE THIS STRUCTURE DOESN'T CARE ABOUT HIMSELF. THIS PERSON HAS LET HIMSELF GO . . . AND HE'LL PROBABLY LET HIS LOVEMAKING GO, TOO."

Enough said?

YOUR HEALTH KICK—FOR HER TOUCHDOWNS

Suppose you were trying to produce a champion racehorse. You gave him all the best training techniques; you let only the best jockeys race him; you saw to it that he got all his immunizations—but you fed him a steady diet of AstroTurf!

Would it be any wonder that he couldn't even finish his heat in the preliminary races?

If you think my suggestion of an AstroTurf diet is ridiculous, how long has it been since you looked at the labels on the food you eat?

"Artificial flavoring, artificial coloring, sodium benzoate added as preservative." Read on! Pretty soon you'll get the idea that you're eating "artificial food." As far as nutrients are concerned, you probably are! I came to the end of my nutritional rope the other day when I picked up a carton labeled "artificial margarine"!

48

Obviously, as we consume more and more foods with additives and preservatives, we owe it to ourselves to beef up our bodies with the nutrients we miss in our diets. After all, if a badly nourished racehorse can't make it through his first heat, how is your badly nourished body going to function when *she* wants you to make your paces through a heat with *her?*

OVERFED AND
UNDERNOURISHED

We Americans are, nutritionists are fond of telling us, the most overfed and undernourished people in the world.

We start at babyhood, consuming prepared baby foods that abound in calories but also abound in artificial preservatives, sugars, and fats. By the time we're pre-teeners, we probably have plenty of money in our jeans—and plenty of corner stores at which to stoke up on carbonated drinks, candy, potato chips, and all the junk foods that put on the pounds without building strong bodies.

Many of the foods we assume are high in nutrition are high only in the *cost* of the small amount of nutrition they contain.

CONVENIENCE FOODS AND
YOUR CONSTITUTION

Consider the frozen packaged dinners. Don't the colorful covers *look* pretty and make the contents *look* healthful?

The shocker is a government announcement that:

1. A man would have to eat *six* "television" dinners a day just to consume his minimum number of calories.

2. Even if the six dinners *were* consumed, the nutrients in those six dinners would still be inadequate.

49

THE TRAINING TABLE

Unless you worked hard at a competitive sport and had a coach who made you maintain strict training, you probably had such bad eating habits by the time you turned twenty that it would take a year to make you stop craving the junk foods that were such a big part of your diet.

If most American men planned their spending and their stock portfolios the same way they planned their diets, they'd be broke and penniless by the time they reached thirty!

You'll begin to notice an immediate difference in your ability to control your weight and in your general well-being after you begin following a diet low in carbohydrates and high in protein.

Remembering what we've already conceded about the health hijackers in the foods we eat and the environment in which we live, how can you best supplement your diet with the vitamins that will bring your health potential up to its highest possible peak—and keep it there?

So many of our ills can be traced to vitamin-poor diets. I hope the day comes when a person can present himself to an analyzer who will, literally, "assay" the properties of his body, determine those areas in which he may be deficient, and prescribe the proper vitamins to correct that deficiency.

If you're confused about vitamins, I'm not surprised! In every newspaper or magazine we pick up, we're assaulted with some leading nutritionist saying that massive doses of vitamin — will cure or prevent ——. Six months later, a research specialist reports that this isn't so! Nutritionists often point an accusing finger at the food industry, for ruining the natural nutrients in our food by use of too many preservatives. The food industry retaliates by claiming that the nutritionists are wrong; that no American need supplement the good food on his table with any vitamin whatsoever.

I, for one, while the debate rages on, am saying that you owe it to yourself to protect your body in the interim period while the medical men, the nutritionists, and the food industry continue their debate. If you allocate just a bit of the money you spend on the "junk" food (potato chips, beer, soft drinks, etc.), you can purchase the same vitamins I take to supplement my diet. You will feel better, look better, and have the assurance that your body is getting everything it needs for better living and better loving.

THE SUPERREGIMEN

I'll share with you the vitamin program and some of the diet tricks that a good friend of mine—a really super he-man whose name is practically a household word—follows every day.

I asked my friend (who always seemed in the pink and ready for any kind of action, despite his grueling television schedule) where he gets all his pep. He laughed and gave me a tour of his enormous kitchen, which boasted a refrigerator just to accommodate all his vitamin supplements and health foods. I was a little surprised at first, but after he explained *what* he took and *why*, I began to understand that he had put more than a little time into learning about nutrition.

I wouldn't dream of prescribing a vitamin program for you, but I'll tell you how my friend supplements his daily three meals. Here's what he takes:

His Daily Vitamin Program

200,000 units vitamin A	(He says it keeps skin healthy, young looking, and blemish free.)
500 mg. vitamin C	(You can't tell him it doesn't prevent colds. He says he used to have a perennial postnasal drip and fell sick whenever the weakest cold germ came his way!)

51

100 mg. pantothenic acid	(He says this keeps the adrenals perking and helps the body resources lick infections.)
2 mg. vitamin B$_2$	(For healthy eyes and skin.)
2 mg. vitamin B$_6$	(An enzyme that helps the body use the protein it takes in.)
½ tsp. inositol	(Related to preventing high-cholesterol problems.)
5 mg. folic acid	(For prevention of anemia.)
5 mcg. biotin	(He says he takes this mainly for his skin, but did you know that rats have been made completely bald when deprived of biotin? Do you get the hint?)
300 mg. para-aminobenzoic acid	(For healthy skin and hair. PABA *ointment* is the most marvelous sun-protection lotion ever invented, even for you fair-haired boys and blondes like me!)
1 tbsp. wheat germ oil	(A wonderful natural source of vitamin E.)
10 tablets 200 units vitamin E	(He proclaims E to be the best vitamin friend his heart could ever have.)

To this daily supplement of vitamins, he adds eight ounces of carrot juice (he makes it fresh in his juicer and drinks it immediately; it's the best late afternoon "cocktail" anyone could ever have).

Every morning he has four to eight ounces of freshly squeezed orange juice, for nature's own way of providing that wonderful vitamin C.

He tries to insure that each meal provides a balanced portion of the elements he needs to stay healthy and full of vitality.

BREAKING YOUR FAST

I'd like to add a word about breakfast. Having lived on a "New York schedule," I think I came to understand why so many of you men rebel against a good, nourishing break-fast. Unless you're living back on the farm, or are fortunate enough to work very close to your home, you probably *don't* eat dinner right at 6:00 P.M. or 6:30 P.M. If you're a commuter, or if you often entertain clients at dinner, you very likely don't actually eat until 10 P.M., perhaps later. Is it any wonder that you're not as starved by 8 A.M. the next morning as the fellow who ate at 6 P.M.? In the case of the 6 P.M. diner, he's gone a full fourteen hours without food by 8 A.M. the next morning. In the case of the late diner, he's gone only eight hours without food by the same time in the morning. Speaking from a standpoint of nutrition, he could probably go until 11 A.M. or noon before having his "breakfast." So don't make a big thing of choking down eggs and toast and juice or whatever comprises your wholesome breakfast—if it isn't time for you to eat. Adjust any diet, any schedule, to your own individual set of needs. Do, however, try to pack thirty protein grams into your first meal of the day, no matter what time you eat it!

MY OWN FLEXIBLE DIET

My schedule is often topsy-turvy, what with television appearances, lecturing, university seminars, winter and summer sports, my exercise program, and my personal life. Here's how I tailor my diet so I can fit in the proper foods for good nutrition with my own vitamin regimen to insure that I operate at peak vitality every single day.

I always have breakfast, but not necessarily at the same time every morning. I have one egg. My favorite companion to the egg is a small chopped sirloin meat patty, which I vary occasionally with two slices of very crisp bacon or a piece of ham steak or some sausage or cottage cheese. About every other day I add one-half slice of 10 percent whole wheat bread. I have four to eight ounces of freshly squeezed orange juice and four ounces of skim milk. I skip the coffee stimulant because I think it's better to do without it, and coffee is very hard on the digestive system, you know.

LIQUID LUNCH: A POWER POTION

I know you executives have your liquid lunches—three Bloody Marys and six cigarettes. And I know you sometime get concerned about your weight and your health and switch to another kind of liquid lunch—one of those canned or powdered diet drink concoctions that fill you fuller of carbohydrates than most milk shakes and leave you hungry half an hour later.

My own liquid lunch is like neither of your liquid lunches.

It's a power-packed pep potion that I borrowed from my health-conscious friend, and it is my favorite midday break. (It's nutritious enough to serve as a full day's meals if you want to blend up a quart and store it in the refrigerator. If you do this, just keep it in your blender jar and reblend for a minute or two before drinking.)

The POWER POTION (one serving):

1	glass skim milk
½	tsp. bone meal
½	tbsp. powdered liver
¼	tsp. magnesium oxide powder
¼	tsp. calcium gluconate
1	tbsp. lecithin
1	tbsp. soy oil
1	tbsp. dry yeast, health formula
1	tbsp. pure vanilla flavoring
3	tbsp. unsweetened pineapple (crushed or chunks)
¼	tsp. nutmeg

FIND YOUR HEALTH FOOD
"PUSHER"

Now I realize you probably don't have these items up in the kitchen cabinet alongside your pancake mix, cornmeal, and enriched flour. You might not even know where to get all the ingredients. Cheer up! Simply take this book over to your nearest health food store and tell the shapely, pretty girl behind the counter what you need.

She'll assemble all the ingredients (even the pineapple) while you browse around. (That trim, affluent man you see in the organic foods section is probably the president of your corporation.)

While you're in the health food store, you might ask about the vitamin supplements I listed in my daily program and stock up on the ones you've decided to use.

You can pack your liquid "power-potion" lunch in a small thermos for the days you know you're going to be deskbound (how much better than sending out for a greasy cheeseburger!) or for the days you're going to be working out at a nearby gym (more about this later).

The pep drink is INSTEAD OF A SOLID LUNCH, not IN ADDITION TO A SOLID LUNCH, if you need to lose weight.

One misinformed man I knew used to drink a whole can of "reducing formula" BEFORE consuming his regular lunch. How ridiculous! He might as well have had a chocolate malt before consuming an additional eight hundred or so calories.

When I'm not able to have my own liquid POWER-POTION lunch, I have ONE item from the list below, plus, perhaps, a crisp salad and skim milk.

1 A ground sirloin patty with cheese (broiled)
2 Cottage cheese (as much as I'm hungry for)
3 1 small can salmon, shrimp, or crabmeat

EASY DOES IT

All of us work under pressure. You men especially do, day in and day out. If I am working under pressure of a busy next-morning schedule that makes me want to shudder as I look at it before turning out the light, I pamper my stomach.

In my case it's my stomach that seems to shout "STOP" when things get too hectic. In your case it might be your lower back, your head, your neck. A supervitamin regimen helps all of us cope with pressure.

I'll give you other ways to cope with pressure in this chapter. Right now, let me tell you how I pamper myself when my stomach is rebelling against the pressure of too many appointments and obligations set within one twenty-four-hour period of a one-week, one-month, or one-year period.

First, I avoid highly seasoned foods—all of them.

Second, I try to eat more mini-meals rather than assault my system with three large ones.

Third, if I suddenly realize I'm in the middle of a long period between meals and am hungry, I resist the urge to give my system "junk" food. Instead, I have a few large spoonfuls of cottage cheese or several ounces of skim milk.

Fourth, if my stomach gets jumpy (as I said, all our systems show the effect of nerves and pressure one way or another), I take two tablespoons of acidophilus (a liquid available at your health food store) in a glass of milk or buttermilk. Acidophilus provides the "good" bacteria our system needs for proper digestion of food. Yogurt that you make yourself (without preservatives) will do the same thing, but the liquid product is more readily available.

Fifth, I stick to my POWER POTION for lunch.

Sixth, I never go to bed, no matter how tired I am, feeling "empty." I have a little milk or some cottage cheese or half a fresh melon (or some other nonacidic fruit).

DINNER—YOUR DOWNFALL?

I've known men, and women, too, who are very careful about their vitamin and calorie intake every hour of the day—before nightfall.

Comes dinner hour—no matter what hour—and the whole program literally goes to pot.

For your vitamin-rich and calorie-wise dinners, choose as I do. Resist the urge to binge at dinner. Actually, it's worse if you do—since after dinner, you're less likely to be exercising or moving, or in any manner burning up calories (unless you count brushing your teeth vigorously or cavorting with her on your private playing field).

If you gorge on high-calorie "junk" at dinner, you're literally pound wise (all day long) and pound foolish (blowing all your hard work in one hour of high calorie living).

My own dinner is very simple.

If guests are coming, it might include prime rib of beef au jus (natural juices). Or chicken, baked to brown perfection. Or fresh fish broiled with tangy lime or lemon slices and the zesty visual accent of paprika.

Complements are a cooked vegetable (my guests love the fresh squash from my own organic garden—I buy from the organic foods section of my supermarket in the winter) and a crisp salad with lush red tomatoes, bell peppers, and other fresh vegetables in season.

My vegetables have *not* been sprayed with DDT. They haven't been injected with artificial coloring. They haven't been dosed with preservatives. They're gorgeous and fresh and nutritious. The meat I eat has a natural red color, not added coloring so "only the butcher knows for sure" what the meat *really* looks like. I eat beef from my Colorado

ranch rather than beef from animals that have been con-
tinually treated with antibiotics or have received "tend-
erizing injections" shortly before slaughtering. (You can
find a butcher who specializes in organic beef, wherever
you live.)

PINK AND PERKY

If all these precautions seem excessive, it's just because
I'm taking the best possible care of the only body I'll ever
have! I want you to do the same because they're not going
to issue you a new body, you know.

I want you to be a superlover.

And there's just no way in the world to make sure that
precious four-to-eight-inch area of your anatomy is going
to be perky and in the pink of condition—unless you nourish all those other inches of your body every day!

Chapter 4
Work Out To Make Out

There comes a time when we just have to be able to tell the boys from the girls (or the men from the women, if you prefer), and that's when we're allllllll alone together under the sheets.

Maybe there's a place for unisex in a mod clothing store, but it's just no fun at *all* in the bedroom. IF YOU BOTH LOOK ALIKE IN BED, ONE OF YOU IS SUPERFLUOUS!

Suppose she has a thirty-eight-inch bosom—and your pathetic chest nearly equals her thirty-eight when it's fully expanded!

Suppose she looks down at your entwined legs and finds that your calves and thighs are so underdeveloped that she can't tell yours from hers!

Worst of all, suppose you have a pot tummy that makes you look six months pregnant—how in the world is she going to feel feminine in such a situation, or should she feel guilty for making love with you—in your delicate condition.

If you're to be a successful lover, she needs to feel your masculinity in every portion of your body. The mere fact that you have a six-inch area (goodness, that's half a foot

—have you ever thought of it that way?) that says "I'm a he" just won't get the job done.

You've just got to BUILD your shoulders and upper arms, BUILD your thighs and calves, BUILD your neck and back muscles, and TRIM your torso and tummy unless you want to have the dubious distinction of being considered a really neuter unisexed partner in bed. The biggest bat and best-inflated balls in the world won't prove your masculinity if you present her first with a body that's as slender and unmuscled as a girl's.

A really great and well-developed body is the first thing about you that a woman notices. It's also the last thing to leave her mind (and physically, it lingers long after your bat and balls have played their final inning and retired for the night!).

MEASURING UP TO MASCULINITY

For every inch of every man there's a measurement he ought to consider ideal if he's to score in the SEX league instead of being left with the girls in the UNISEX league. Find your height on the masculinity chart and see how each area of your body measures up. (Expand your chest, flex your arm and leg muscles, and hold in that abdomen before submitting your sexy self to the tape measure.)

NSL NUMBER ONE DRAFT CHOICE
(National Sex League)

Height	Neck	Biceps*	Calves	Chest*	Waist*	Thighs	Weight
5′3″	14⅜	14⅜	14⅜	39¾	27⅞	21½	125–131
5′4″	14½	14½	14½	40⅛	28	21⅞	126–136
5′5″	14⅞	14⅞	14⅞	40⅝	28⅝	22⅛	136–143
5′6″	15	15	15	41½	29¼	22⅝	141–151
5′7″	15¼	15¼	15¼	41⅞	29⅝	22⅞	146–156
5′8″	15⅞	15⅞	15⅞	42½	30	23⅜	154–164
5′9″	15¾	15¾	15¾	42¾	30½	23½	160–170
5′10″	15⅞	15⅞	15⅞	43¾	30⅞	23⅞	168–178
5′11″	16⅛	16⅛	16⅛	44⅛	31⅜	24⅛	174–184
6′	16½	16½	16½	45	32	24⅞	185–195
6′1″	16¾	16¾	16¾	45⅝	32½	25	192–202
6′2″	17	17	17	46	32⅞	25¾	198–208
6′3″	17	17	17	46⅞	33¼	25⅞	207–217
6′4″	17⅜	17⅜	17⅜	47⅛	33¾	26	216–226

* Chest measurement is expanded, biceps are flexed, waist is pulled in.

NSL NUMBER TWO DRAFT CHOICE
(National Sex League)

Height	Neck	Biceps*	Calves	Chest*	Waist*	Thighs	Weight
5′3″	13⅞	13⅜	13⅜	37¾	28⅝	20½	115–124
5′4″	14	13½	13½	38⅛	29½	20⅞	118–125
5′5″	14⅜	13⅞	13⅞	38⅝	30½	21⅛	128–135
5′6″	14½	14	14	39½	30¾	21⅝	130–140
5′7″	14¾	14¼	14¼	39⅞	31⅛	21⅞	135–145
5′8″	15⅛	14⅝	14⅝	40½	31½	22⅜	144–154
5′9″	15¼	14¾	14¾	40¾	32	22½	150–160
5′10″	15⅜	14⅞	14⅞	41¾	32⅜	22⅜	158–168
5′11″	15⅝	15⅛	15⅛	42⅛	32⅞	23⅛	164–174
6′	16	15½	15½	43	33½	23⅞	175–185
6′1″	16¼	15¾	15¾	43⅝	34	24	182–192
6′2″	16½	16	16	44	34⅜	24¾	188–198
6′3″	16½	16	16	44⅞	34¾	24⅞	193–207
6′4″	16⅞	16⅜	16⅜	45⅛	35½	25	208–215

* Chest measurement is expanded, biceps are flexed, waist is pulled in.

NSL NUMBER THREE DRAFT CHOICE
(National Sex League)

Height	Neck	Biceps*	Calves	Chest*	Waist*	Thighs	Weight
5′3″	13⅜	12⅜	12⅜	35¾	29⅞	19½	109–114
5′4″	13½	12½	12½	36⅛	31	19⅞	112–117
5′5″	13⅞	12⅞	12⅞	36⅝	32	20⅛	121–127
5′6″	14	13	13	37½	32¼	20⅝	122–129
5′7″	14	13¼	13¼	37⅞	32⅝	20⅞	124–134
5′8″	14⅝	13⅝	13⅝	38½	33	21⅜	133–143
5′9″	14¾	13¾	13¾	38¾	33½	21½	139–149
5′10″	14⅞	13⅞	13⅞	39¾	33⅞	21⅜	147–157
5′11″	15⅛	14⅛	14⅛	40⅛	34⅜	22⅛	153–163
6′	15½	14½	14½	41	35	22⅞	164–174
6′1″	15¾	14¾	14¾	41⅝	35½	23	171–181
6′2″	16	15	15	42	35⅞	23¾	177–187
6′3″	16	15	15	42⅞	36¼	23⅞	182–192
6′4″	16⅜	15⅜	15⅜	43⅛	37	24	192–202

* Chest measurement is expanded, biceps are flexed, waist is pulled in.

NSL NUMBER FOUR DRAFT CHOICE
(National Sex League)

Height	Neck	Biceps*	Calves	Chest*	Waist*	Thighs	Weight
5′3″	12⅞	11⅜	11⅜	33¾	31⅜	18½	103–108
5′4″	13	11½	11½	34⅛	32½	18⅞	109–114
5′5″	13⅜	11⅞	11⅞	34⅝	33½	19⅛	116–120
5′6″	13½	12	12	35½	33¾	19⅝	117–121
5′7″	13¾	12¼	12¼	35⅞	34⅛	19⅞	113–123
5′8″	14⅛	12⅝	12⅝	36½	34½	20⅜	122–132
5′9″	14¼	12¾	12¾	36¾	35	20½	128–138
5′10″	14⅜	12⅞	12⅞	37¾	35⅜	20⅜	136–146
5′11″	14⅝	13⅛	13⅛	38½	35⅞	21½	142–152
6′	15	13½	13½	39	36½	21⅞	153–163
6′1″	15¼	13¾	13¾	37⅝	37	22	160–170
6′2″	15½	14	14	40	37⅜	22¾	166–176
6′3″	15½	14	14	40⅞	37¾	22⅞	171–181
6′4″	15⅞	14⅜	14⅜	41⅛	38½		

* Chest measurement is expanded, biceps are flexed, waist is pulled in.

TESTING YOUR PRESENT
CONDITION
Can You Rate With the Pros?

A man who has never experienced true physical fitness has never known the pure joy of functioning at full capacity for life and love. He may have made love many, many times, but he has never known the joy or the phenomenon of making love with all the vitality, vigor, and masculinity that can be his.

Before you get your semiannual physical checkup, I wouldn't want you to do anything more strenuous than just holding this book! Dr. Kenneth Cooper, author of the best-selling book *Aerobics*, told me that on a recent television show he put two stars on a treadmill test—and was thoroughly alarmed when he discovered that to continue the treadmill might have seriously overtaxed their hearts! He stopped them immediately, of course, but we had quite a discussion about prescribing exercises for anyone who has not had a complete physical examination.

Visit your doctor. Show him the tests below before you take them.

Then, armed with your clean bill of health, see how you rate. Are you a Reject, an Amateur, a Semipro, or a PRO?

I am asking you to do ONE of the three tests for Endurance and Stamina. You may choose among Running, Stationary Running, or Swimming for your Endurance-Stamina.

In addition, do the best for Strength and see how you rate.

TEST FOR ENDURANCE AND
STAMINA AND STRENGTH

Since Jack Benny and many other men want to be and stay between the ages of thirty and thirty-nine, these

65

standards are to test not your chronological age, but your physiological age.

REJECT	1 mile or less in	12 minutes
AMATEUR	1 to 1.20 miles in	12 minutes
SEMIPRO	1.20 to 1.50 miles in	12 minutes
PROFESSIONAL	1.51 miles and over in	12 minutes

TEST FOR STATIONARY RUNNING

REJECT	75 steps per minute for (5 or 6 inches off floor, lifting feet)	1 minute
AMATEUR	75 steps per minute for	5 minutes
SEMIPRO	85 steps per minute for	10 minutes
PROFESSIONAL	95 steps per minute for	15 minutes

TEST FOR SWIMMING

REJECT	50 yards or less	2½ minutes
AMATEUR	200 yards	5 minutes
SEMIPRO	400 yards	8 minutes
PROFESSIONAL	900 yards	19 minutes

TEST FOR STRENGTH

REJECT	Bench press a 60-lb. barbell	30 times
AMATEUR	Bench press a 100-lb. barbell	30 times
SEMIPRO	Bench press a 125-lb. barbell	30 times
PROFESSIONAL	Bench press your body weight	30 times

I hope you were able to rate yourself right along with the pros by doing the proper tests within the twelve-to-nineteen-minute time limits. If you rate with the pros, it's a good indication that your stamina and endurance and vitality are such that your lovemaking can go on and on and on, without any regard at all for the clock!

BUILDING YOUR MAKE-OUT
MACHINE

YOUR HE-MAN HEADQUARTERS

For a relatively small investment, you can become a superstar in the masculinity market—without even leaving your own den. You can even keep soft music going while you lift and push your way into a marvelous new body!

You can, of course, join a health spa or your "Y" health club. But why not start out our program at home so that you know you can always get to the equipment you need to do the exercises that mold your body into new manliness?

You should see the elaborate exercise equipment that Johnny Carson keeps in his apartment. His body shows that he uses it.

You can get everything you need from Diversified Products, the largest manufacturers of exercise equipment in the United States. Their equipment is carried in almost every sporting goods and department store across the country.

BEAUTIFUL BODIES ON SPECIAL!—WHAT YOU'LL NEED

Item	Cost
100-pound set of SuperBells (you may add more weight as you progress)	$24.95 (approx.)
Multipurpose incline squat bench	$50.00 (approx.)
Door bar gym	$3.00 to $6.00

For less than eighty-five dollars, then, you've got a lifetime gym that you can use in the convenience of your own home. Later, should you get lonely (but I doubt that this will happen, since she'll be so enamored of your new body that you'll hardly have the time to get out), you can move your exercise regimen to a spa, health club, or "Y."

EXERCISES FOR MAKING OUT

SEX IS SOMETIMES A MATTER
OF SHOULDERING YOUR
RESPONSIBILITIES

Military Press

Purpose   To increase the size of the shoulders, plus triceps and chest (this is one of the best three-in-one exercises)

Positions
1   Stand with feet about 12 inches apart.
2   Hold a 40-pound barbell close to chest, shoulder-length apart.
3   Press up until arms are completely straight.
4   Lower to chest level and repeat.

BEST WAY TO HER BREAST IS
WITH A BIG CHEST

Bench Press

Purpose   To build chest; will also build triceps

Positions
1   Lie on bench, hold a 60-pound barbell at chest level, shoulder-length apart.
2   Push weight up until arms are straight.
3   Lower to chest and repeat.

MAN WITH STRONG FOREARM IS
SELDOM LIMP WRISTED

Forearm Flexes

Purpose   To increase the forearms

Positions
1   Stand with feet apart, elbows bent at waist.
2   Hold a 10-pound dumbbell in right hand and bend wrist down.
3   Bring wrist back level with arm, and repeat repetitions with right hand before transferring to left.

ONLY DUMBBELLS LET ARMS
STAY SKINNY

Arm Curl
    Purpose   To build biceps
    Positions   1   Start with a 15-pound dumbbell in your right hand, arm extended downward beside your right leg.
                  2   Bend your elbow and raise weight toward your right shoulder.
                  3   Repeat all repetitions on right hand before repeating exercise with left hand.

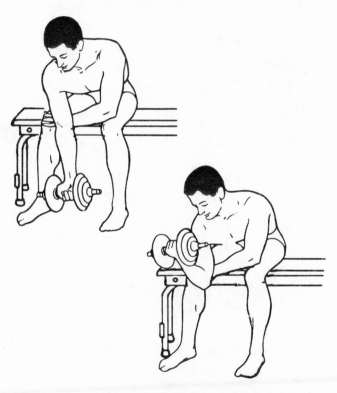

Arm Curl: Positions One (top) and Two (COURTESY DIVERSIFIED PRODUCTS CORP.)

HE WHO DROPS HER AT THRESHOLD WILL NOT MAKE IT TO BEDROOM

Triceps Extension

Purpose   To increase the triceps as well as to strengthen shoulders

Positions   1   Stand with feet apart for balance.
            2   Hold a 40-pound barbell with palms turned outward, elbows bent.
            3   Elbows shoulder level, extend hands straight forward (arms will be parallel to floor).
            4   Return to position 2 and repeat.

Triceps Extension: Positions Two (top) and Three   (COURTESY DIVERSIFIED PRODUCTS CORP.)

MAN WITH SCRAWNY LEGS
FINDS SELF IN LOVER'S LIMBO

Calf Raise

Purpose   To add muscle and shape to your calves

Positions
1. Stand with feet about 18 inches apart, hold a 30-pound barbell on your shoulders, feet turned slightly outward.
2. Bend knees.
3. With knees still bent, raise up on toes.
4. Stay up on your toes as you straighten legs.
5. Lower heels to floor.

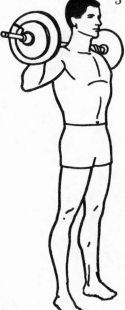

Calf Raise: Positions One and Two (top);
Position Four   (COURTESY DIVERSIFIED
PRODUCTS CORP.)

FIRM THIGHS GIVE PROMISE OF
MORE FIRMNESS IN AREA
BETWEEN

Half-Squat

Purpose	To increase the size of thighs
Positions	1 Stand with feet 12 inches apart, a 10-pound barbell on shoulders.
	2 Keep back straight as you bend knees to a half squat.
	3 Return to standing position.

Caution: Do not do this exercise if you have back trouble. Do not try to go over halfway down—it is too hard on the knees.

Half-Squat: Positions One and Two
(COURTESY DIVERSIFIED PRODUCTS CORP.)

WEAK BACK—NO GOOD IN SACK

Chinning Bar
Purpose   To increase the size of your dorsal muscles
Positions   1 Hold chinning bar with hands about 24 inches apart.
             2 Pull body upward until chinning bar is in back of neck.
Repetitions Start with position 1

Caution: If behind the neck it too hard for you, reverse and pull up with chinning bar in front and switch as soon as you get enough strength.

SHE WHO FLOUNDERS IN YOUR FLAB WILL NOT GET LOST— JUST MISLAID

Waist Twist
Purpose   To reduce the waist
Positions   1 Stand with feet 18 inches apart, hands on hips.
             2 Lean slightly backward and twist to right side.
             3 Touch both hands to floor on right side.
             4 Come back to center and lean slightly backward and twist to left side and touch floor on left side; repeat twenty times.
Repetitions Start with twenty (ten each side); add five repetitions a week to each side until you have reached your goal.

MALE MUSCLES FEEL GOOD IN TUSSLES

Sit-Ups:
Purpose   To reduce and firm upper adomen
Positions   1 Lie on back with knees bent, feet hooked under couch or barbells.

73

2 Clasp hands behind neck, sit up, touch head to knees.
3 Repeat ten times first day; add one repetition a day for the first week; add 3 repetitions a week until you reach one hundred repetitions.

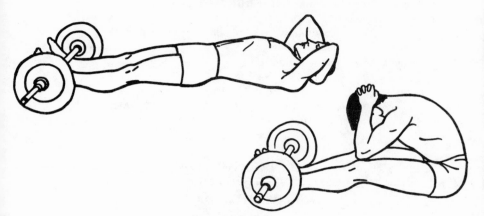

Sit-Ups: Positions One (top) and Two
(COURTESY DIVERSIFIED PRODUCTS CORP.)

UNLESS TUMMY'S FLAT—SHE
CAN'T GET TO YOUR BAT

Leg Raise

Purpose   To flatten, reduce, and firm waist and lower abdomen

Positions   1 Lie on back on floor or table.
2 Raise both legs straight up, then bend into chest.
3 Lower legs in the same manner.
4 Start with ten; add one a day for first week; after first week, add three repetitions a week until you reach one hundred.

74

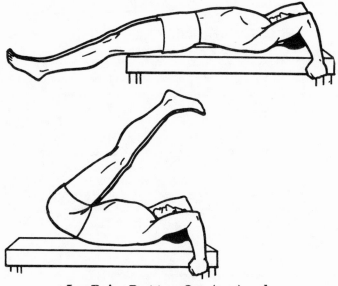

Leg Raise: Positions One (top) and Two (COURTESY DIVERSIFIED PRODUCTS CORP.)

1st week	2nd week	3rd week	4th week
1 set of 7 repetitions	2 sets of 8 repetitions	3 sets of 9 repetitions	3 sets of 10 repetitions

ADD FIVE POUNDS TO ALL EXERCISES

5th week	6th week	7th week	8th week
3 sets of 7 repetitions	3 sets of 8 repetitions	3 sets of 9 repetitions	3 sets of 10 repetitions

Rule 1   It is important to rest about 10 seconds between each set of repetitions. For example, for 3 sets of 10 you would do 10 repetitions, rest 10 seconds, do 10 more repetitions, rest 10 seconds, and then do the last 10 repetitions.

Rule 2   Do *all* sets of the same exercise before moving on to the next exercise.

Rule 3   When you reach 3 sets of 10, add five pounds of weight to your exercise and reduce your repetitions to 3 sets of 7. Work your way back up to 3 sets of 10. Continue this procedure until you have reached the goal I've set for you.

Rule 4   Rest at least one minute between each exercise. For example, do the Military Press, then rest before you do the Bench Press.

Rule 5   All the body-building exercises should be done three time a week with a day in between. (Sit-ups and Leg Raises are not body building, but reducing and firming, exercises.)

DOING IT AT THE OFFICE

How can you expect to do it with her after work if you haven't done it with yourself at the office?

Just sitting at a desk all day can completely undo you as a lover. That's why I'm suggesting "doing it at your desk" so you can "do it with her after work." One "doing" helps the other "doing," you know. You can perform your love-making with lots more vitality and vigor if you're not a sluggish blob from eight hours of sitting in your executive suite (including one hour sitting in your favorite restaurant).

Sexability sags after sitting that long. Muscles get lax, and your playing equipment gets more and more ready for retirement with each passing sedentary day. You owe it to her to keep your physical structure perking, and you can do it right at your own desk. If you're the boss, no one can take issue with you; and if you're not the boss, you can surely explain that you're just "doing it" for the sake of love rather than taking a dull old coffee break like all the other fellows.

Here's how you can do it at the office so you can do it better after work. Doing all these exercises will take only five minutes, so try them at your morning and afternoon coffee break, and before you go to lunch. That's three "doings" at the office, the better to equip your body for three "doings" with her after dinner. What a lovely dessert!

NECKING, EXECUTIVE STYLE

Necking, executive style, will keep down the tightness in the neck that's caused by tension.

1. Place your hands on your forehead, with your elbows extended to the sides.
2. While resisting firmly with your hands, attempt to push your head forward. Push as hard as possible; rest; and repeat for a total of fifteen seconds.

WITH YOUR BACK TO THE WALL

1. Sit in your chair or stand against a blank wall.
2. Raise the arms to 90 degrees, palms down until the backs of the hands touch the doorway.
3. Keep your arms raised to 90 degrees and your stomach muscles tight.
4. Push your arms backward, pinching your shoulder blades together, push for six seconds.
5. Repeat three times.

THE NINE-TO-FIVE NECKER

1. Sit in your chair.
2. Grasp your chair alongside your buttocks.
3. Attempt to shrug your shoulders, maintaining hold on your chair as though you were trying to lift it.
4. Pull for six seconds; relax; repeat three times.

77

THE PENCIL PUSHER'S PINCH

1 Sit in your chair at your desk.
2 Put your hands behind your head, fingers laced.
3 Bring your elbows forward and touch them together in front of your face.
4 Pull your elbows backward as far as you can, pinching your shoulder blades together. Push for six seconds.
5 Keep your tummy muscles tight and your back straight.

THE BOSS'S BACK BUILDER

1 Raise your feet and legs off the floor.
2 Place your hands on your chair, and pull your abdominal muscles in.
3 Slowly try to raise the buttocks off the chair. Hold for six seconds and slowly lower the buttocks to the chair. Repeat six times.

THE CONFERENCE TABLE
COMPRESS

1 Sit at your conference table or desk.
2 Attempt to compress the desk by pulling your arms together, resisting for six seconds. Repeat three times.
3 Exercise at different positions on the desk to work the muscles differently.

A COFFEE BREAK FOR YOUR
LOWER BACK

Straight leg raising backward

1 Lean over the desk; slowly raise one leg as high as possible.

78

2   Hold for six seconds. Slowly return to the starting position. Repeat two more times.
3   Repeat using other leg.

TRIMMING THE CORPORATE TUMMY

1   Sit in a chair and hold the sides of the chair for balance.
2   Slowly straighten the knees, raising the legs as high as possible.
3   Hold for six seconds; relax; repeat two more times.

YOUR STRONG BACK IS VERY GOOD IN THE SACK

Women love to feel your rippling muscles and your firm, masculine power as they snuggle with you in bed. They love your broad shoulders, your slim, sensuous hips, the strong ridge of muscles on either side of your vertebrae.

But women, understandably, are turned off by a weak back—especially a weak back with an owner who constantly complains about it.

"Oh, my aching back." What a lovesong! If you're encased in a back brace, how romantic!

For those who claim man was never intended to walk upright in the first place, a "bad back" is a natural progression, a result of aging.

For those of you who intend to keep thrilling her in bed, never suffering a twinge of back pain, there are PREVENTIVE exercises that will insure that the wonderful backside of your splendid body will never become a liability in bed!

THE DROOP THAT REFRESHES

Sit on a hard chair, erect, with arms hanging loosely in front of you. Relax completely, and let your body go forward and limp until your head is down between your knees. Count to three, then tighten your stomach muscles as you pull yourself upright into a sitting position. Relax, and repeat the head-to-knee droop. Repeat five times.

THE HIGH-STEPPING LOVER

This one is like doing a bent-knee high step, except that you can lie flat on your back to do it! Lie with your arms stretched in back of your head, and legs bent at knee. Bend right knee and bring it as close to your chest as possible, then reextend leg. Pause. Repeat with other leg. Repeat ten times.

THE DOUBLE WHAMMY

This time, bring knees up to your chest and hug them close to your chest with your clasped arms. Hold the position for a count of 5, then reextend legs. Pause. Rest. Repeat five times.

FLAT ON YOUR BACK!

Lie again with your knees bent, feet flat on floor. Place arms over head. Now, press your back as flat as you can against the floor, by firming up the muscles in your lower abdomen and in your lower back. Place your hand so you can be sure the backbone is flat against the floor, with no space between the back and the floor. Repeat five times.

EXERCISES FOR THE ARTISTIC LOVER

Department store mannequins (in case you've never peeked at one in a total state of undress left in the window by a careless display director) are *stacked!* They've got bumps in the right places, and they're slim where they ought to be slim.

Yet, even if they came equipped with certain female equipment, how would you like to make love to a dummy? It would be no fun at all—because she wouldn't MOVE in bed.

MOVING and FITTING your bodies together is as much fun as the actual game itself!

You're as badly off as a department store dummy if you don't learn to move your entire body to turn her on, and to fit your body into every little niche of hers. Stiff lovers never get a perfect fit.

ARTISTIC LOVERS learn smooth, controlled, sensuous body movements that make them wows for the game of love. If you practice these movements, you'll find you can incorporate them in your general posture, giving a hint of the kind of lover you really are—and proving you're no dummy when it comes to lovemaking.

THE PRESS THAT IMPRESSES HER

Pelvic Action One: Stand erect with your feet together, and with your toes and chest pressed against a wall. Try to touch the wall with your pelvic area. Now relax the pelvis, and gently thrust the pelvis forward against the wall. Relax and thrust twenty times. Work up to forty times.

AIMING YOUR THRUST

Pelvic Action Two: Now place your feet wide apart, and repeat the gentle pelvic thrusts against the wall. Can't you see how this will help you in lovemaking?

ARE YOU HIP TO LOVE?

Pelvic Action Three: Stand as though you were going to take a racing dive off the side of the pool, with

81

your knees bent and your arms extended behind your back. Now, knees still bent, execute the pelvic thrusts. Still holding the position, move your hips slowly from right to left. Repeat this twenty times, and work up to 40. Are you getting the swing of it?

HURRAY FOR THE HULA!

Pelvic Action Four: Do you remember the hula hoop? Imagine that you are being screened in slow motion while doing the hula hoop. Feet apart—knees bent, rotate your hips s-l-o-w-l-y, around and around. Then change directions. Repeat twenty times, then increase to forty.

HIP AND TUCK WITH LOVE

Pelvic Action Five: Still standing, with your feet apart and your knees slightly bent, tuck your hips under as far as possible. Now, push your hips slightly forward and to the right side. Hold two seconds. Come back, and repeat on left side, holding two seconds. Repeat on right, up to twenty times. Be rhythmic.

ONWARD AND UPWARD!

Pelvic Action Six: Lie down, with your legs slightly apart, your knees bent, and your arms relaxed along the sides of your body. Now, raise your buttocks off the floor, slowly thrusting the pelvis forward. Move slightly to the left, then lower buttocks. Hold two seconds. Now raise buttocks and slowly thrust pelvis forward and slightly to the right. Repeat twenty times on each side, keeping a smooth rhythm.

SHOULDERING INTO A CUDDLE

If you watch a football player—shall we take Joe Namath?—moving down a football field, you'll notice a definite grace of movement. It's this lithe, animal grace that turns women on more than anything else.

Work on these shoulder movements and your whole body will move with new grace—the better to land on *her* soft shoulder!

Shoulder Roll One: Stand with your feet slightly apart, and let your arms hang at the sides of your body. Now, slowly bring both shoulders forward, up, back, and down. You will be making small circles with each shoulder simultaneously (almost as though you were doing the swimming crawl stroke without involving your arms).

Shoulder Roll Two: Stand with your feet apart and your arms at the sides again. Roll the right shoulder forward, up, back, and down; the left, repeating each side twenty times.

IMPROVING ADAM'S RIBS

Rib Cage One: Stand erect, and raise your chest up and forward. Slowly circle to your right, back, and to your left. Repeat twenty times on each side.

Rib Cage Two: Stand with your feet apart, and your arms down at your sides. Extend your rib cage to the right, keeping your hips perfectly immobile. Slowly move your rib cage to the left. Repeat twenty times, and work up to forty times.

KEEPING THAT TORCH LIGHTED

There may come a day when you think you just can't fit exercise into your schedule. Perhaps you can't make it

83

to the "Y," or to your athletic club, or to your health spa, and you're not in the mood for a full workout with your home gym, but you can alter your rigid daily routine occasionally and still get your exercise accomplished.

Investment man and race-car driver Jack Eiteljorg has a stationary bicycle positioned right in front of his television set. He sets the tension and goes at it for thirty minutes worth of viewing (including commercials, when he's not snacking the way some men I know are). By the end of the thirty minutes, he has worked up a good sweat and feels much better than his sedentary friends.

THE LIVELY LUNCHEON

Two businessmen I know often meet for lunch . . . but it's not a typical two-hour-with-cocktails affair at all. Instead, they get together for a quickie lunch (sometimes just an apple that they bring to work), then stroll all over the city while getting their business dealings attended to in style.

TRAVELING MAN'S DELIGHT

I bought an out-of-shape man I know a dandy jump rope to carry in his luggage with him. He started out jumping only one minute a day—and worked up to a robust twenty minutes jumping time. He said that he learned to watch television movies while jumping and that it wasn't bad at all—reminded him of the early jiggling of the television picture with every passing airplane! He added push-ups and sit-ups to round out his regimen.

FUN IN BIG HOTELS

I traveled a great deal with one of my business associates who was as interested in staying in condition as I was.

We found that if we walked upstairs several times a day (my favorite room in New York is on the Hilton's twenty-fourth floor), we felt better and had a lot more pep, too! (We stopped at every sixth or seventh landing.) From his performance on the stairs, I could tell why his wife was so very happy with their marriage. Every hotel has stairs—so you can try this, too! Who knows whom you might meet on the landings!

YOUR CIRCULATORY
STIMULATORS

To give your Make-Out Machine that you build by weight lifting the go-go-go power that makes you really healthy with plenty of staying in power, you need a series of exercises that I call "circulatory stimulators."

I wouldn't want you to build a beautiful, muscular body without building the circulatory system that insures you greater health and go-power!

Your Make-Out Machine needs to be built in two ways.

First, you need the handsome, virile, muscular body that makes her take a second look at you as a love partner. That's the OUTER MAN.

Second, you need to build the lung capacity and the heart and circulatory system that make these vital organs operate at their peak efficiency for every physical need (including lovemaking). This is the INNER MAN.

One set of fitness exercises builds only half the man I want you to be if you're to be a better lover longer. Both sets will build you to the peak of TOTAL MASCULINITY FOR TOTAL FITNESS AND TOTAL SEXUALITY.

Your system will benefit when you are able to master my circulatory stimulators. I've been advocating such exercises on my television shows and in my exercise studios for some years. The "aerobic," or "Air Force," exercises embody many of these ideas, and your use of the circulatory stimulators, month by month, will build your body into a pro-body to go with pro-lovemaking. As with my other exercises, it's necessary that you get a complete physical checkup before you start (including that all-important exercise electrocardiogram). Tell your doctor that you intend to start the following exercises. Get his advice, but I recommend a second exercise electrocardiogram about six weeks into your circulatory stimulator program.

WHAT CAN YOU EXPECT?

With the circulatory stimulators, you can expect an easing of tension, a gradual lessening of daily fatigue, a building of lung power and endurance, and a fine toning of muscles and reduction of flabbiness.

Circulatory stimulators are designed to make you increase your lung capacity. Most of us use only one-fourth of our lung capacity in our daily breathing in and breathing out! For total fitness you must increase this capacity by performing *one* of the following exercises every day.

I am giving you the rates at which you should be able to start to work out every day, providing you are pronounced healthy by your doctor and are between ages thirty and forty. You may adjust these exercises upward or downward if you're above forty or below thirty.

YOUR CIRCULATORY
STIMULATORS
Pick and Choose

Cycling: You should be able to pedal two miles in sixteen minutes or less. Over a two-month period, work to increase your daily cycling to four miles in thirty minutes or less, cycling daily.

Swimming: You should be able to swim 100 yards in four minutes or less. Build your swimming workouts up to a daily workout of 400 yards in fifteen minutes or less over a two-month period.

Stationary Running: You should be able to run at a rate

of seventy-five steps per minute, lifting your feet six inches from the floor with each step. Begin by running in place for one minute daily. Over a two-month period, increase to five minutes daily.

Running: You should be able to run one mile in twelve minutes. In a two-month period, increase your distance to two miles and run it in twenty-four minutes.

All these "stimulators" make your muscles work, causing them to burn oxygen and release carbon dioxide. As this happens, your nerve centers tell your lungs to function better, and deeper breathing takes place, giving the body a greater supply of oxygen. These stimulators thus help not only your circulatory system but your pulmonary system as well, since they cause your lungs to work up to their full capacity (instead of that lazy, lazy one-fourth capacity that nonexercises often use).

You will notice that by their very nature the "stimulators" are the sort of exercises that cause you to breathe in a rhythmic in-and-out pattern. These exercises aren't the sort that cause you to put forth sudden spurts of action, interspersed by pauses. You might find that sort of exercise, say, in football (You're quarterbacking and returning the kickoff. You make a ten-yard dash, zigging in and out. You're stopped suddenly. You wait during the huddle and the snapback, then you dash again.)

Instead, these exercises keep your system at a steadily stepped-up pace, with the regularity you need to build your lung capacity and increase your circulation in a manner that's kind to your system. (By the time you work up to a maximum performance in one of the areas of exercise and stay there, you'll find your body is better at delivering quick spurts of energy, as well as long-term performance.)

THAT INNER SATISFACTION

After eight weeks of working at your choice of the circulatory stimulators, you will have taken a giant stride forward toward developing the INNER MAN that matters so very much as far as your long-term health is concerned.

Keep up the stimulators along with your regular exercise program. She may never see your heart, your lungs, or your amazing network of arteries, veins, and capillaries. But she'll FEEL the difference in your lovemaking if you've done your homework and kept the inner man hale and hearty with my daily stimulators.

POINTERS FROM A PRO

The major leaguers whom you admire don't just fall into their athletic ability, you know. Aside from their strenuous workouts in their particular sports, they put in a lot of time building up their general physical stamina.

I asked Jerry Hill, fullback for the World Champion Baltimore Colts, what daily exercise routine he followed. The six-foot, 212-pound Jerry (who nailed down a cool $15,000 in the 1970 Super Bowl) gave me a detailed outline of his exercise regimen. If this seems like a lot of exercise, just remember that the following are mere warmups before actual practice starts!

WHAT'S A WORKOUT?

Here's how Jerry Hill and the Colts are doing it, and for them, all this is just in a day's work (out).

1 One run around the football field.
2 20–25 push-ups.
3 20–25 sit-ups.
4 (One or two minutes worth of neck-bends. Jerry goes backward on all fours plus his neck, and "walks" around. He says it's great for strengthening the neck.)
5 The Grass Drill. The one is for agility. Players run in place; then suddenly hit the ground at the sound of a whistle; then spring to their feet and run in place again.

6 The Crab. This is great for agility and quickness. Bend forward, with knees slightly bent, hands on ground. Scoot forward and back, then to right, then to left, scuttling on all fours like a crab.

7 The Tire Jump. Old tires are lined up in a double row. Jerry gets a running start and hits the hole on the left tire with his left foot, then the hole on the right with his right foot, until he hits the end of the row. Then he repeats the exercise going backward. This is wonderful for developing both agility and those thigh muscles. (If this seems easy, you must be using the tiny tires from your daughter's baby buggy. No fair! Try full-size automobile tires, and just see what a workout you get.)

8 The Squat and Shuffle. Alternating his right and his left foot in a kickout like a Russian dance, Jerry uses this exercise to keep his leg mucles in A-1 condition. Keep this one up for two or three minutes and see how your muscles tighten up and how hard your lungs work.

9 The "Gassers." This is running forward the entire width of the football field at top speed. Pause ten seconds; then run backward at top speed. Jerry uses the "gassers" up and down a hill in his neighborhood park and says they're a great conditioner. (Jerry bought his suburban Denver home just because it was right across the street from a large park with plenty of running space.)

Are you trying to keep up with Jerry's routine?

If so, you've earned a tiny rest. Lie down on the cool grass and try these exercises, continuing Jerry Hill's regimen:

10 The Bicycle. Lying on the grass, roll back on your shoulders, support your hips with your hands, and "bicycle" vigorously through the air for three minutes.

11 The Leg Raises. Lie flat on your back, with your arms flat at your sides to use as bracers. Lift both legs up, up, up, over your head until they

touch the grass in back of your head. Repeat ten times.

12 The Scissors Kick. Still lying on your back, raise your legs six inches off the ground and spread your legs apart. Bring your legs together, and lower to grass. Repeat twelve times.

13 Arm Circles. Stand erect (if you're able) with feet apart. Extend arms out to sides, parallel to ground. Keep arms outstretched, and move in twelve-inch circles.

14 Chest Expanders. Stand with hands at waist, feet apart. Fill chest up and out, and lean backward slightly. Repeat at least ten times.

15 Leg Stretch. Lie on your stomach, and extend arms out to your sides. Kick your right leg up, then your left. Repeat ten times for each leg.

16 Leg Lifts. Lie on your back, extend arms out to sides. Kick your right leg up to ninety-degree angle. Hold five seconds. Repeat with left leg. Repeat ten times for each leg.

THE HEALTH KICK TICKET

I'd like to see every American man working out at his neighborhood park the way handsome, fit Jerry Hill does.

And I'd like to see neighborhood parks within walking distance of every residential area. I'd like to see more parks downtown, for that matter. I'd also like to see long, long networks of bike paths so that both children and adults would have a safe way to "exercise" their way to and from school or work.

If businesses like Kodak, U.S. Steel, and Xerox are putting gymnasiums in all their new buildings so they can keep their executives in tip-top health, why can't every company follow suit?

Washington has made some progress by putting black-bordered warnings on cigarette packs. Why can't they legislate against the cigarette completely? Why can't they legislate for compulsory fitness programs in all public schools—just as they've legislated for integration in all schools?

I'd like to run for president on a HEALTH ticket. Then, after my party got all the programs started, perhaps in another generation, all men would look and be as healthy and full of vitality as star athletes like Jerry Hill.

WHERE THE VIPS GET
THEIR VIGOR

When it comes to exercise, I've found that Very Important People have Very Impressive Programs.

I've found that the more successful men have worked out exercise and diet programs that keep them in demand for corporate promotions.

Corporations want men who are lean and fit and healthy.

These successful men are aware that:
1 BEING OVERWEIGHT CAN COST AN EX-ECUTIVE ABOUT $1,000 PER POUND PER YEAR IN WAGES.
2 ONLY ONE IN TEN TOP MEN IN TOP JOBS (those paying over $25,000 annually) WAS MORE THEN TEN POUNDS OVERWEIGHT.
3 ALMOST FOUR IN TEN MEN IN LESSER JOBS (those paying less than $20,000 per year) ARE OVERWEIGHT.

There's a direct link here; corporations want to project a vigorous image—and fat and flabby executives detract from that image.

Voters (especially in these days when candidates get so much television exposure) want a congressman who physically represents the American ideal. For that matter, if you were going to choose between two doctors, all other factors being equal, wouldn't you select the one who was tanned and slim and full of pep—whose own body made him look as though he practiced what he preached?

Men at the top of every field stay in top physical condition.

It's a funny thing—the more packed their schedules seem to be, the more they seem to be able to make the time for a good exercise program. I've found it's only those flabby workers stuck in second-rate jobs who complain to me, "I'm too pooped after work to do much besides watch television." Most of them just won't believe me when I tell them that a brisk jog around several blocks would relax their tensions and build up their energy level much more than the drinks they consume for a quick lifter to help them "unwind."

Just listen to my roster of some of the VIPs who make exercise and proper diet a part of each day's routine:

Mickey Rooney starts off each day in his Fort Lauderdale home with a half-hour session of weight lifting, chinning, and push-ups. For Jack Kelly (with a sister like Her Serene Highness, Grace, a fellow has to keep in shape), the fitness program includes running, swimming, squash, weight lifting. He works out four days each week, an hour each day. Warren Cowan, president of Rogers, Cowan, and Brenner, Inc., the well-known public relations firm, keeps his public image fit and trim by working out with dumbbells and doing exercises.

ABC's West Coast publicity manager, Joe Maggio, at age forty-seven, has the physique of a man in his twenties. He stays that way by jumping rope five hundred to one thousand times daily, taking daily walks, striding up hills, running on the beach, and swimming.

Senator William Proxmire (D., Wisconsin) is a familiar sight in Washington; he jogs almost five miles from his home to the Capitol every day and often swims three dozen laps of the Hilton Racquet Club pool to keep in superb physical condition.

For President Richard M. Nixon, bowling and jogging constitute a keep-fit program (with regrets that his schedule doesn't allow more time for working out).

In Lorain, Ohio, *Journal* editor Irving Leibowitz has been getting rave notices from his fellow journalists for his "new look." He gave up lunches for swimming eighteen laps each noon and nibbled a small ground meat pattie instead of consuming a lot of calories. Result? He's down from 196 to 153 and says his love life has changed from dreary to "great"!

Colorado's handsome Governor John Love keeps his good-looking physique by horseback riding and skiing.

From the "Today" show studio, Joe Garagiola told me he's a great believer in stretching exercises and that he works out with his family for thirty minutes twice a week. "Whenever I press an elevator button I draw in my stomach muscles and tighten my rump—and hold it until the elevator shows up," Joe said. "I do the same thing sitting in a car and waiting for the lights to change. This adds up to fifteen or twenty isometric exercises a day." He indulges in the luxury of pasta only once a week and sticks mostly to meat the rest of the time. He uses his basement gym with its bicycle, punching bag, and rowing machine regularly, rides a bicycle, uses an outside gym twice a week, and nixes drinking to relax in favor of using his home sauna before bedtime. Whew!

The *Atlanta Journal's* Furman Bisher must cause quite a stir when he's staying in a hotel! Ordinarily he starts his day by jogging for twenty minutes around his Atlanta home. But when he's on a trip, here's what he does. "I run across hotel furniture to the bathroom—anywhere to keep moving around the room. I imagine patterns and move back and forth in an X. I've done this in one way or another for over ten years; it makes me work up a good sweat, then, after I take a good shower, I feel like going out and tackling somebody!" Furman also golfs but agrees with me that it's not too beneficial as an exercise unless he carries his own clubs as he walks around the course.

What makes executives run? Their desire for fitness, that's what! Joggers are everywhere.

California's Senator Alan Cranston holds the world record for fifty-five-year-olds in the 100-yard dash and starts out each day running at 6 A.M. for a good workout before tackling his office duties at 9 o'clock. He competes in meets all over the world.

In Denver, dentist Francis McCabe started a running-jogging program with his area YMCA—and parlayed his running skills into ribbons at regional track meets. In Dayton, National Cash Register President Stan Laing is a big jogging fan, while in Atlanta, WSB–TV Manager Elliot Heald is a fan of mini-jogging and combines this exercise with a low-carbohydrate diet to keep in shape. Dayton *Daily News* editor Jim Fain uses his "Y" track for jogging in the winter when he can't use the tennis courts and supplements the jogging with a daily routine that consists of twenty-five leg raises, sit-ups, push-ups and squats. He plays squash twice a week in the winter as well.

In the capital city, other dedicated joggers include two husband-and-wife teams. They're former Attorney General Ramsay Clark and Mrs. Clark, and Senator Mark Hatfield (R., Oregon) and Mrs. Hatfield. There is also Strom Thurmond, married to a woman forty-five years younger than he, and, at sixty-nine, father of a baby girl!

Bill Cosby enjoys a good game of basketball, which bounces off a lot of calories per hour; Art Linkletter combines handball and swimming to keep in shape.

San Diego syndicated columnist Donald Freeman has an isometric trick similar to Joe Garagiola's tummy tucks during elevator and red-light waits. Don makes it a habit to suck in his tummy every time the phone rings! If you get a number of calls every day, think how you could firm your waistline! (And how much better off you'd be than the fellows who light up a cigarette every time the phone jangles. Don also strides a mile a day and nixes drinking and carbohydrates.)

David Rimmel, who's Sunday editor for the *Cleveland Plain Dealer*, catches up on the news by tuning in the "Today" show while he works out doing back bends, waist twists, and running in place. With a wife who's a nurse and a daughter who's a dietician, things are pretty healthful around the Rimmel household, and he lost eight pounds last year with a sensible diet and his daily morning workout.

Down in Texas, *Fort Worth Press* columnist Jack Gordon gets a walk of at least three miles after lunch each day. Houston's Jack Harris, president and manager of KPRC–TV, chairman of the NBC Affiliates Committee, and president, Maximum Services Telecast, beats him by walking *six* miles a day and using both wheel exercisers and a commercial exercise machine. And multimillionaire H. L. Hunt, now eighty-three, has maintained excellent health all his long life. He doesn't get all his exercise just by counting his money, either. He regularly practices yoga, has an exercise teacher, and spurns drinking, smoking, coffee, tea, and carbonated beverages. When I dined with him, we had bread baked from fresh grains ground in his home and enjoyed a delicious (and nutritious) carrot cake for dessert.

Tennis, which burns up a whopping 420 calories per hour, is a favorite sport of such VIPs as Robert S. McNamara, U.S. Ambassador to the United Nations George Bush, Senator George McGovern (D., South Dakota), New York City Mayor John Lindsay, and Bishop Fulton Sheen.

Jazz dancing (and is that ever an effective workout) attracts Tom Poston; he and I have gotten our daily exercise together doing jazz a number of times. For Bob Hope, a favorite exercise is walking, walking, walking—a wonderful way to keep fit. Father Francis Bakewell, who boasts the physique of a man twenty years younger, combines jogging with a daily program of weight lifting.

Johnny Carson, whose physique always looks like a million dollars, uses his beautifully equipped home gym. He works out regularly with weights and does stretching exercises, both to limber up and to unwind before going to bed. For John Wayne, the exercise is horseback riding—what else? Handsome actor David Hartman gets his exercise working out with the San Francisco Giants, no less. George Plimpton has taught us all how an amateur feels when he's competing with pros in every sport!

For Wendell Phillippi, managing editor of the Indianapolis *News*, his busy day always includes a Columbia Club workout with swimming, cycling, and weight lifting, and a steam bath as dessert. *Chicago Today* editor Dick Haney makes it a point whenever possible to walk his appointments instead of taking cabs. (Most of you in larger cities would probably save time *and* money *and* your physiques by doing this.)

And FOB James, who's president of Diversified Products, the largest exercise equipment firm in the United States, is a fine salesman for his own lines. His physique testifies to the fact that he does a daily workout with his own weights, dumbbells, and jump ropes. He adds tennis, push-ups, and jogging to this routine—and insures that he'll never lose the build he had when he was an all-American football star for Alabama.

Elston Brooks, amusements editor of the Fort Worth *Star Telegram*, told me he gets his exercise mowing the lawn! Just as I found myself hoping he had a hand mower and cut his grass every day, he sighed and confessed, "Maybe that's why I'm not as good a lover in the winter." He also told me he walked to his car every day and then, in a column, told his readers, "But I forgot to tell Debbie I have an attached garage." No fair, Elston!

I could go on for pages and pages about men who work out to make out . . . in life, love, and business. You see them in increasing numbers every day, because the image of the "successful" American male is certainly changing.

He's no longer the man with the big paunch (the better to display his gold key chain).

The look for the seventies is sexy, slim, sensuous, and sinewy—and the way the VIPs keep the look is by charting part of every day for working out to make out—in every way.

These men live life to its fullest. Not one of them is the sort whose obituary will read:

BURIED AT AGE 75
DIED PHYSICALLY AT 25

TOWARD GLOWING GOOD HEALTH
ALL AROUND THE GLOBE

In Holland I was very much taken with the glowing red cheeks, the healthy skin, and the generally fit and happy attitude of the people. It didn't take me long to pinpoint why the people looked so wonderful. The answer was simple. Adults seemed to cycle everywhere they went. Housewives biked to the store; men biked to their places of work. Children biked to school (not one school bus did I see, carrying pale-faced, sedentary youngsters to a sedentary day of studies). In the winter the people took to their skates to get where they wanted to go. It was a scene right out of *Hans Brinker,* and the rosy cheeks were just as bright and healthy as my childhood storybook depicted.

ALL ABOARD FOR EXERCISE

In the television coverage of President Nixon's historical visit to Peking, I was amazed to learn that *everyone* takes two fifteen-minute exercise breaks a day and that railroad conductors even engage their passengers in a quarter-hour of calisthenics morning and afternoon to relieve the travel "blahs."

I watched small children double-stepping it to their grammar schools, singing and swinging their arms in what in China is called a "living school bus." I hear that they get a half-hour of physical education instruction every day in school, plus a *three-hour* physical culture session every week. Some students are sent to special physical education schools, which turn out the superb athletes we saw doing acrobatics. With such a fitness program, is it any wonder that the new Chinese seem to wear so well? I didn't share the commentator's surprise that "almost all the national leaders are well up into their seventies." It seemed only natural to me that bodies so well cared for would repay their owners with good health through what we Americans tend to consider old age, almost "over-the-hill" age.

THE U.S.A. IS CATCHING ON

Education and industry are beginning to catch on to the need for a beefed-up program of physical fitness in this country.

At the University of Denver, Dr. Larry Gettman is luring professors away from the libraries and into the gymnasium—at least for a little while each day. Profs who want to develop pro bodies participate in thirty-minute exercise classes five times a week to clear the cobwebs from their circulatory system and keep those brains perking at top tilt.

Before being assigned a custom-made program to follow, each professor takes an electrocardiogram before, during, and after stress (using an exercise bicycle). Unless they're already getting regular exercise, they are not allowed to work to full capacity (and their pulse rate is not allowed beyond 150).

Instead, they are given exercises that keep them work-

98

ing for thirty minutes at pulse rates of 120 to 130 beats per minute. (If they're not getting up to a pulse rate of 120, they're not considered to be getting a workout good enough to benefit their cardiovascular system and their lung capacity.)

The profs are given a caliper test at the waistline, back, and upper arms, and an anthropometric at the shoulders, chest, hips, upper arms, wrists, elbows, ankles, and knees, to see how they compare with the norms. (A fit male should have no more than 16 percent body fat on his physique.)

Fifteen minutes of running is alternated with fifteen minutes of calisthenics.

Is the program working? Just consider the case of psychology professor Dr. S., who in one academic year of workouts managed to reduce his exercise pulse rate from a soaring 172 to a far more healthy 128. His cholesterol level has dropped from 240 to 148. He's an enthusiastic exerciser and has made his remarkable progress despite the fact that a back injury makes it necessary that he substitute the exercise bike for some of the more strenuous calisthenics.

Results of the program are being carefully recorded at the University of Denver and at other universities. I hope the results cause an exercise revolution!

Big business is going in for exercise in a big way, too, but I almost hate to tell you why!

Chapter 5
Improve The Package—
Improve Your Chances

When you walk into a crowded room, do eyes turn toward you with admiration?

Do women look at you with desire?

Do men look at you with envy?

Are your clothes perfect for you and for the occasion?

Does your body send out strong messages of warmth, friendliness, virility, intelligence, and confidence?

Do your eyes, your hair, your mouth, your teeth, your profile, and every part of your body make you attractive to others?

Only when you answer yes to all these questions will you know you possess the "packaging" that makes a woman want to select you as a lover, as a mate.

Your "packaging" sells you before she finds out that you're a millionaire or an M.D. or a famous sportsman. It's the first thing she notices about you—and unless the packaging is good, she's not going to wait to discover that what's underneath your packaging is really superb!

If Madison Avenue spends billions of dollars a year improving packaging to lure customers to a certain product at the point of purchase, you should certainly spend just a tiny fraction of that amount perfecting your own packaging. It's one way you can be sure you won't be left sitting alone on the shelf of love after you've watched all the better-packaged products being snapped up by women!

COFFEE, TEA, OR THEE?

I suppose you all know that a stewardess has more on her pretty little mind than coffee, tea, or milk.

And if you think it's just all those gorgeous male passengers, you're on a mad flight of fancy! Oh, she thinks about you, all right, but do you know *what* she thinks?

She doesn't miss one thing about you; these girls are really alert—and they do more male-watching than women in any other profession. No—that's not quite true—the stews see more of you with your clothes on, and doctors and nurses see how you look minus your "fig leaves"!

STEWING OVER MY PASSENGERS
By Marlene

Denver to Baltimore, Flight 174
4:10 P.M., EDT

There were twelve first-class passengers on this flight. All were men, with the exception of one woman who was traveling with her husband. I didn't include some of the men in this list because they were just such nothings that I had no reaction to them at all.

MY NEGATIVE REACTION	*MY POSITIVE REACTION*
Seat 2B He didn't cover his mouth when he yawned while talking to me. This was bad enough—but what a mouthful of old fillings. Ugh!	

MY NEGATIVE REACTION   MY POSITIVE REACTION

He was too flashily
dressed, even though
he was rather ex-
pensively dressed. I
hate big rings on
men!

Oh, the well-worn
lines I have to put up
with. 2B actually said,
"How would you like
for me to take you
away from all this?"
(Come, now!)

He had to show his
power and influence
immediately. He was
presumptuous enough
to say he wanted to
take me to Africa
next February (with
no encouragement at
all from me—and then
he started bragging
that he could get us
a place at the same
game preserve that
William Holden
liked). He's really a
creep with his ad-
vances. No subtlety
at all. He's so eager
that there's no
challenge, even if a
girl were at all
interested.

Seat 3B	This one was a United crew member. He was ultraconservative in	Not demanding as a passenger.

103

MY NEGATIVE REACTION	*MY POSITIVE REACTION*

every way. He had
sort of a "little boy"
appeal, but that was
all.

Seat 5D   He played with his
food and kept sucking
a lime wedge (very
ugly). He was over-
weight and had a
paunch, but he was
wearing a mustard-
colored velour shirt
which just accentu-
ated his overweight
state.

Seat 6A   The paunch most men
seem to have is 6A's
problem to some
degree. He has a
small mole on his left
cheek which should
be removed. Puzzling
—what was he trying
to tell me? He delib-
erately worked it
into our conversation
that his *wife* was
French!

His greeting was very warm
and friendly, and not pushy.
He was seemingly genuine.
He was well dressed and
had on a pale blue shirt
(my favorite on men) with
a navy blue and yellow
polka-dotted tie—nice and a
little unique. He had on
black horn-rim glasses, and
his face was slightly tanned.
He had a pleasant facial
expression in repose or while
reading. He always said
"thank you" and seemed to
notice me, not just as a
"servant." He smoked a nice
pipe, and I found him very
masculine. He was reading
during almost the entire
flight, which gave me the
impression that he was self-
contained (as opposed to
2B, who stared at me all the

MY NEGATIVE REACTION   *MY POSITIVE REACTION*

time as I worked and couldn't seem to entertain himself).

Seat 7A   He was too old to be interesting, and he looked like a dried-up spider. His appearance gave no sense of grace. He did something very repulsive to me, and frequently done by male passengers— he "waves his commands." It didn't occur to him to have enough consideration to *ask* to have his plate removed.

Seat 7C   He had a very swarthy appearance, with shifty eyes, and he seemed very nervous. He seemed to relax after a couple of martinis.

Seat 7D   He was well dressed (tan suit with tangerine shirt) and was clean cut.

Baltimore to Denver, Flight 177
6:20 P.M. EDT

This flight carried six male passengers and three female passengers.

MY NEGATIVE REACTION   MY POSITIVE REACTION

Seat 1B He was wearing a
white undershirt be-
neath a white shirt;
the top part of the
undershirt showed
beneath his open
collar. He had on
short socks, but may-
be that was OK be-
cause he was such a
short man. He was
very shy and unas-
sertive.

Seat 2B He was wearing
brown loafers with a
relatively nice busi-
ness suit. He did not
say "thank you" once.
He had such a bad
overbite and buck
teeth that he should
invest in braces. He
had a pot tummy and
wore a narrow tie.

Seat 4B A paunch (another
one!). He wore non-
descript glasses. He
wore dull clothes and
had a rather toadlike
appearance.

Seat 5B He was too casually
dressed for first-class
travel. He had on a
blue Hawaiian print
shirt. His pants were
baggy . . . men want
to see our "behinds,"
and I think they

MY NEGATIVE REACTION   MY POSITIVE REACTION

should return the
compliment and give
us a formfit view, too.

Seat 6B   He stood in front of
me talking, but I can't
remember what he
said because he was
combing his hair
simultaneously. What
a turnoff!

Now, Marlene is a smart cookie. She may have made all
these comments to help me—but I know that like all other
females, she would keep very silent about each passenger's shortcomings if she ended up with one of them in a
more personal relationship than stewardess-passenger.

If she had wanted to be perfectly blunt, can't you just
imagine how her flights would go? The loudspeaker flicks
on, and you hear:

"Good evening, ladies and gentlemen. Welcome to
Cloudco's Superjet Service to Los Angeles. We will be
serving cocktails and dinner shortly. Ground temperature
in Los Angeles is a pleasant 72 degrees, and we expect to
arrive on schedule. I am Miss Corbin; working with me is
Miss Whitten. We will do everything possible to insure
that your flight is a pleasant one.

"Now, passenger 18, your garlic breath is awful. Where
in the world did you have lunch, and don't you think a
breath mint is in order before you start talking to that
lovely girl beside you?

"Passenger 2C, where in the world did you get that
narrow-lapel suit? Most of our first-class passengers shop
more at places like Eric Ross than Salvation Army.

"3A, do you have to scratch your scalp like that? Is what
you've got communicable? Should I arrange for quaran-
tine in L.A.?

"3D, you slouched so badly as you approached the plane that I thought you might be in need of a wheelchair to make it. And I see you're not over thirty. Straighten up!

"4C, I had you on my last flight, and I swear, you can smack while eating, even with your mouth closed.

"4D, are you in love with your own feet or something? The way you keep your eyes downcast makes you look sullen and shifty. You're not another hijacker, are you?

"6A, the back of your head makes me want to cry. There you've got a nice head of hair except for that bald spot the size of a fifty-cent piece. Looks like someone took a big bite out of your scalp! Haven't you heard of mini-toupees to cover spots like that?

"6D, I haven't seen a pair of clear plastic eyeglass frames in YEARS. Were they your father's or something?

"7B, we're not allowed to wear sunglasses on our flights —and that outfit of yours is going to blind me. Did you have to pull out every single hot-colored item in your closet and put them all together at the same time? I could understand it only if you were carrying a white cane!

"7C, how much coffee did you have to drink, and how many cigarettes did you have to smoke to get your teeth so yellow? You definitely need a professional cleaning to remove all those pitted areas that make up the yellow. And while you're at it, get those two front teeth shortened by filing. I'm tempted to toss you a carrot because those choppers make you look just like Bugs Bunny.

"7D, you have the raspiest voice I've ever heard. You look so nice—until you open your mouth. Can't you take a throat lozenge or some voice lessons—or is it possible that you're in the last stages of some terrible disease?

"8A, the way you walk makes you look like a pregnant duck. I know your feet aren't webbed, so why do you waddle like that? Men who pigeon-toe-waddle or walk as if they just got off a horse are enough to turn off any woman.

"8D, your animation is out of control! I don't know who

108

that man you're talking to is, but your squeaky voice and high nervous laugh is going to totally unnerve him before we get halfway to L.A. Enthusiasm is fine, but you're simply frenetic.

"And 9A, will you PLEASE STOP OGLING Miss Whitten? There's just no way she's going to let you make a pass at her—especially when you've obviously had too much to drink before boarding.

"Thank you very much for your attention, ladies and gentlemen; I do hope you enjoy your flight, and thank you for flying Cloudco."

Ginnie, another stewardess I know, rather wistfully told me that she'd started out as a stewardess in eager anticipation of being around interesting, well-traveled males. After five years as a stewardess, observing men (and their faults) at such close range, she is rather badly disillusioned. Ginnie contributed the following list of her male passengers' bad manners:

1 CLIPPING NAILS IN PUBLIC.
2 MAKING FUN OF STEWARDESSES. She says she gets sick of having to defend her job with some "smart guy" cracking well-worn jokes on every flight.
3 BEING IMPOSSIBLE TO PLEASE. Ginnie cited a not-so-fond memory of a man who asked her for "just a little salad dressing." She asked, "On the side?" He said no. She put what she considered a small amount directly on his salad. He bellowed out, *"Don't you know what a little means?"* What terrible behavior.
4 REFUSING TO SHOW GRATITUDE—EVEN FOR EXTRA SPECIAL SERVICE. Ginnie remembers the time she found a man's briefcase left under his seat. Concerned that it might contain papers he needed for a business meeting (it was a morning flight), she took the briefcase and

dashed up and down Dulles Airport (quite a workout) until she found the briefcase's owner. Breathless, she passed him his briefcase. He looked at it, puzzled *"Huh? Oh yeah. . ."* was all he said before strolling away, briefcase in hand. Ginnie had to work like mad on her next flight to make up for the fifteen minutes she had lost in tracing the man. An she didn't even have the memory of his "thank you" for all her trouble.

5 THE "OH, YOU KID" APPROACH. Ginnie says there's one of these jokers on almost every flight. He knows everything about stewardesses, and he knows they're loose because he's made out with so many of them, so why doesn't she just relax and enjoy what's going to transpire between them. What transpires, his behavior guarantees, is little more than the passing of his dinner tray from her hands to his!

6 ACTING LIKE A FOUR-YEAR-OLD ABOUT FOOD. Their mothers would have spanked them if they'd behaved like this at the dinner table. Ginnie says she's had great big grown men make the following comments:
(On being served anchovies topping a salad.) *"Hell, no, I can't stand the smelly things!"*
(On being served caviar, no less, with cocktails.) *"What's this awful-looking stuff?"*
(On being served domestic wine, before even tasting it.) *"Take this yuk away. I thought I was paying for first-class service."* "This 'yuk,'" Ginnie says, "may not have been grown in France—but its great-grandfather grape was the same that wine grown in France's Beaujolais region had! Men who put down domestic wines ought to be put down for their own lack of knowledge." She'll take a naïve but nice passenger over the blowhard any day, like the man who smiled up at her and said, *"I guess the white wine would go better with my chicken, wouldn't it?"*

7 WAVING SILVERWARE IN THE AIR WHILE TALKING. Ginnie said she lives in

fear of being vaccinated with a fork if the plane hit a sudden dip.

8   TUCKING A NAPKIN IN AT THE COLLAR OR AT THE BELT. "I always wanted to ask such passengers if they wanted me to bring them a high chair," Ginnie purred.

9   USING REALLY LARGE AND FRANTIC HAND EXPRESSIONS WHILE TALKING WITH OTHER PASSENGERS. This, Ginnie thinks, is the mark of an insecure man, one who just can't trust his words and the thought behind them to be worth much unless he punctuates his conversation with gestures.

10   USING HANDS FOR OTHER NO-NOs. As hard as I find this to believe, Ginnie tells me that men have been found masturbating on several of her flights. "Once, just before landing in Chicago, I was going down the aisle, retrieving pillows. I don't see how this man could have helped but see me coming. I smiled, reached down, and picked up the pillow—and there he was, playing with himself! I plopped the pillow back down and just fled to the back of the plane That was one situation they never taught me to handle in stewardess school. Now I approach every male passenger's pillow as though there might be a rattlesnake lurking underneath it!"

ROUNDING IT UP

As further evidence of how much you're in the eyes of others, listen to further comments from stewardesses I've talked to:

"Suit bags with the hangers inside look so much nicer. This can be part of a man's total picture. I saw a garment bag in leather the other day, and it really looked elegant. Men who are careful about their dress hang their bags up as soon as they board, so the suits don't get wrinkled."

"Men who care how they look take off their coats and ask me to hang them up right away so they won't look rumpled. They never toss them up into the overhead racks —although I've seen men do this."

"We call attaché cases 'brain bags,' and they tell a lot about a man. Successful, well-organized men are very productive on flights. They're usually very busy . . . and usually very well groomed. Other men use their 'brain bags' for some rather bizarre things. I've seen an old pair of socks, or some of our miniature liquor bottles, or even golf balls inside some of the briefcases. It's a sure sign the men aren't too busy."

"About the two-drink liquor limit . . . I think if a man is sitting next to where we're serving and asks very, very politely, we'd often relax a rule. Of course, if he's in the first row of seats, and we'd have to walk past all the passengers to serve him—or if he's had too much to drink—it would be out of the question."

"I've had passengers demand to be served their dinners early. Again, it's all in how it's done. If he's sitting near where we start to serve, it makes it easier. I think if a man handled his request right, we'd try to help. Say, if a man told me when he got on, 'I've hardly had a bite to eat all day,' and asked where he could sit to get served first, we'd do everything possible to see that he was given the first tray."

"One of the most thoughtless things men do is stick their feet out into the aisles, which are crowded enough to begin with. It makes it almost impossible for us to serve. Arms, too—they shouldn't hang out into the aisles."

"I can't believe that men eat the way they do. They hunch over—and smacking, smacking, smacking of food seems to follow me up and down the aisle everytime I've served a meal. I even see men with full mouths turning to smile and talk to their companion in the next seat. Between the noise pollution of the smacking and the visual pollution of the mouth full of food, I wonder how these guys ever made it far enough in the world to have the money or the job with expense account to pay for the plane ticket."

"The foul language some passengers use is just unbelievable. I can understand a good curse word if a man drops a hammer on his toe or something, but to be foulmouthed about an underdone steak is simply unforgivable. I'd report him to his boss if I could. No woman in any job should have to put up with bad language."

"Some men have an instant kind of rapport if they want to ask us for a date. They seem to know just how to go about it, and to start conversation with something that sort of binds us together. A neighbor of mine in Chicago started off our relationship by carrying two loads of laundry up to my apartment for me. It sounds silly, but by the time he'd done this, I knew at least two things about him: One—he cared. Two—he was a good conversationalist. I don't like to be just asked point-blank for a date. Any topic of conversation that eases us into a relationship helps . . . even if it's a decoy, like carrying my laundry."

"No girl minds flattery. For instance, I wouldn't take a call from a man I didn't know and accept a date. But if he said, 'I remember you from my flight from Detroit and realized we lived in the same apartment. I got your name and number from the doorman, and I'd really like to get to know you better.' This approach was so direct and so honest and so positive that I went out with the man several times and like him very much."

HOW TO GET DEEPER PENETRATION

I can tell you how to get deeper penetration with a woman from the first instant you lay eyes on her. You can even do it in a room crowded with people and, if you're not too obvious about it, right under her husband's nose!

You get the first sort of penetration (before you make a perfectly horrible scene at your next cocktail party) with your EYES.

Your beginning step in developing sex appeal must be

to develop your eye appeal. When you learn how to use your eyes in a sexy way, to develop bedroom instead of living room eyes, you've increased your total sex appeal by a good 50 percent.

Not all eyes look alike.

But one thing all SEXY eyes have in common is FEELING.

That's why I say that on your very first meeting with a woman you're attracted to you can achieve deeper penetration. When you have feeling and caring and interest and admiration in your eyes, you can penetrate to the depths of her desire. As you look into her eyes for the first time, you can send her the very strong signal that "I am interested in YOU. I am interested in everything you have to say. You have my complete and undivided attention, and, if you're willing, I'm going to show you a lot more, and a lot more intimate ways that I have of showing you that I think you're really marvelous."

FIND YOUR TYPE

Some men have eyes that can almost send out electric sparks. They exude energy, the bright-eyed look of health, and vitality. When you have these eyes, and they light up when you look into HER eyes, she gets the feeling that she was the one who turned you on. She's almost yours already!

Other men have eyes that just seem to smile . . . to project a contented twinkle that tells a woman, "I'm delighted with my role as a male, and I'm ever so much more delighted now that I've found a female like you to make my maleness complete."

Still other men have eyes that are rather dreamy looking and slow moving . . . seemingly with little energy behind them. When eyes such as these concentrate on a woman and give their undivided attention, they can really turn a woman on so much she's almost spellbound;

114

hopefully, before too long, those same eyes will have begun the scene that gets her bedroom bound!

EYES IN THE KNOW

Politicians have learned how to use eye language with maximum results. When I was at the Presidential Inauguration in 1969, I had the opportunity of meeting many state governors. Not one failed to pass the eye test. They meet and greet many people each day, especially when they are campaigning. They often have only a few precious seconds to make a favorable impression or win votes. Along with the firm handshake and flashing smile, they establish eye contact almost immediately. Although the encounter may be extremely brief, they make the most of it. Most show business personalities also use eye language to project sex appeal.

One of my friends, the president of a highly successful public relations firm, has been happily married for eight years. She told me that recently as she was getting undressed for bed, she caught sight of her husband in a full-length mirror. He was watching her disrobe with eyes full of admiration, tenderness, and desire. He didn't have to say a word, but what he said with his eyes made her feel beautiful, feminine, and wanted. She joined him in bed and repaid his unspoken compliment tenfold.

Movie critic Dorothy Manners recently interveiwed Omar Sharif. She was as captivated by his eyes as are most movie fans. She said they were more overpowering when you met him face-to-face than fan magazines or even motion pictures indicate. "He gives you his undivided attention . . . looks at you with sincere interest . . . not out of the corner of his eyes like most men one meets."

I was similarly impressed with Omar's eyes when I was on the Irv Kupcinet show with him. As I talked to him I had the feeling that I was the only woman in the world at that moment. He rivets you with his eyes, and you find yourself spellbound under his gaze.

Another actor who makes great use of eye contact is Stephen Boyd. He has a deep, penetrating quality that is unforgettable. Although I met him rather briefly, he gave me a message with his eyes I'll never forget. He makes you feel all-woman, which I am sure accounts to a great degree for his success in pictures and in his personal relationships.

THE EYES HAVE IT!

There are many businessmen who learn how to use their eyes effectively. When they want a favor like getting a rush job or a letter out, they can look at a woman employee with a soft, beseeching look that's hard to resist. More often than not, they will get the desired result, and often without running into overtime pay. On the other hand, when the situation calls for it, a businessman can adopt an icy, imperious look that lets the employee know who's boss in no uncertain terms.

According to Julius Fast, in his best-selling book *Body Language*, "Of all parts of the human body that are used to transmit information, the eyes are the most important and can transmit the most subtle nuances." Stop and think for a moment of all the adjectives used to describe the language of the eyes: piercing, steely, dreamy, bedroom, sloe, knowing, starry, wise, hot, melting, sensuous, insistent, and so forth. How many times have you heard such expressions as "he undressed me with his eyes," "he had bedroom eyes," "his eyes took command of the room," "I melted under his gaze," "when he looked at me my knees started to quiver."

If you're still doubtful about how important eye language can be, take the time to watch two women as they talk to each other. Women make far more effective use of their eyes than men do. You'll notice that as they speak, they are not paying as much attention to what is actually being *said* as they do to each other's eyes. If you're ever driving behind a woman driver who has a woman beside her, watch how frequently the driver will take her eyes off the road to look at her passenger as she speaks. A woman can tell whether the other woman is being truthful not so much by what she says but by her eyes. They are a dead giveaway!

The eyes have been called "the mirror of the soul" and for good reason. If a person is "shifty eyed" or deliberately avoids eye contact, he will not be trusted. Sometimes it will cost him a job opportunity because he will not look the interviewer in the eye.

A friend of mine uses his eyes more effectively than anyone I know. When he meets a girl for the first time he immediately establishes eye contact. He will catch her eye and hold it for a few seconds. While the introduction is under way, he will give her a penetrating look and then after a few seconds look away. If she has held his gaze without looking away first, he's made a point.

Great writers throughout our literature have been well aware of the importance of the eyes. Your great-grandfather used to read . . . "He fixed his gaze upon her tenderly. Their looks met, lingered . . . then she melted into his eyes." The chapter always ended there leaving all the more fun details to the minds of the readers. But, oh, the fun great-granddad had just imagining what went on during and after the melting process . . . why, it might have been the very inspiration for the conception of your grandfather—so where would you have been without the eyes that turned on the heroine that turned on your ancestor?

HOW TO BE A CUT ABOVE
THE REST

If your barber doesn't wield a razor for anything except a shave, you're seeing the wrong man!

What a pair of talented fingers can do with a razor is the

117

difference between letting your hair just lie there on the top of your head—or encouraging it to do what you and your barber decide it should be doing.

And what should it be doing? It should be growing fuller. It should be shaped into sideburns that are long, smooth, and natural looking, not blocked off sharply like two playing zones on an athletic field. If your neck is rather thick, your barber can diminish the thickness by tapering the hair gracefully at the back of your neck. If your neck is too thin, he can make it appear more hefty by cutting the hair more bluntly in a wider area at the back of your neck.

I hope you've kept up with all the innovations in men's hair styling because they'll allow you to help achieve the heads-up look she'll love.

1 *The holding sprays.* Especially as your hair becomes accustomed to falling into new lines, you need a good holding spray to encourage it. These fine sprays keep your hair from flying in all directions, but without the greasy, slick look that turns women off.

2 *Hair coloring for men.* Almost every fellow who's turning gray should get into the coloring game. I was amazed to see an old television friend in Florida last Christmas. He appeared to have shed some ten or twelve years. The difference was his hair—which his stylist transformed from a blah gray to a warm, rich brown. It not only looked completely natural but made his skin look a lot younger, too. Coloring is now a matter of as little as fifteen minutes if you use color that is combined with shampoo. You simply lather it into your hair, wait the prescribed length of time, then shampoo out. You can ask your stylist to start you on a color. Then, if money's a problem, you can purchase the same color at your drugstore or beauty supply store and enjoy your younger look for less money. Incidentally, if you don't like the

first color you use, shop around. You'll eventually find just the right shade to complement your skin tones.

3 *Streaks of color.* I've never known a woman yet who didn't flip over the sight of the sun glinting on a man's blond-streaked hair. It's a very masculine look—very yacht-y or ski-y, but of course you can pull it off only if you're blond toned to begin with.

4 *The electric vibrator.* I'll tell you about some other fascinating uses for the vibrator later on. For right now, suffice it to know that you can do a lot of splendid scalp stimulation with your vibrator. You can bring your entire scalp alive with a four-minute session, with the vibrator strapped to the top of your hand and your fingers helping with the work. Since your scalp runs right into your handsome face, what you do to to maintain a healthy scalp will improve your facial tone and texture as well.

5 *The blow comb and the blow brush.* These ought to be as much a part of your masculinity equipment as your toothbrush. You don't need both, but you do need one or the other to keep your hair in line. The blow comb is designed for more curly hair, and will help you in guiding your hair into the proper look. The blow brush is for fuller, thicker hair. Either device will dry and train your hair after your shampoo.

I could go on and on about how women love your hair. I want especially to mention, however, that there's no need to go around looking ten years older and twenty years less attractive than your potential, just because of baldness. You've probably seen a few before-and-after pictures of men who kicked the bald look in favor of a hairpiece or a hair-weaving job. If so, they were probably tiny pictures in the newspaper. I'd like to have you go to the best men's hair stylist in your city and ask to see his

119

scrapbook of before-and-after photographs. You will be absolutely amazed at the difference in appearance in the men who were smart enough to make an investment in their future by having either a hairpiece made or a hair-weaving job done.

Here are some facts and figures on both procedures:

For a hairpiece, it's much better to get one at the beginning of baldness. That way your public will never need to know that you're even wearing a hairpiece. If you wait, the transformation will be so drastic that people may find it hard to forget that you're wearing one. You'll probably spend about three hundred dollars for a hairpiece. Your stylist will match your hair color and texture perfectly and will make a mold of the area to be covered to insure you of perfect fit. You'll want to buy the best hairpiece you can afford—preferably one made with 100 percent human hair blended to match yours.

While machine-made hairpieces are less expensive, they do not hold up as well. If you can afford to do so, buy two hairpieces and alternate their use. You'll more than double the wear you get out of them.

Hair weaving is an excellent procedure which adds hair to your existing hair (which means, of course, that you'll have to have at least some hair over your ears and around the back of your head so that the stylist has some of your hair to attach the weaving *to!*). You can expect to pay from about three hundred dollars and up for the weaving, styling, and fitting. Once it's on, it's a very natural look—but you'll have to have the weaving adjusted about every other month. The beauty of the weaving is that it stays on your head. It gets windblown when your natural hair does, but you need have no fear of "losing" it. If you're careful, you can even wash and shampoo your new weaving job as you would your own hair.

There's a third way you can improve your appearance by improving your hairline. That's with a hair implant—a surgical procedure that removes skin grafts from portions of the scalp with ample hair and transplants the graft to the bald areas. (This is sort of a Robin Hood philosophy —a stealing from the rich and giving to the poor proposition.) You'll pay your doctor from five to twenty-five dollars for each graft, and you may require up to three hundred grafts to cover your baldness, about fifteen grafts per appointment. The procedure's expensive, but in about three months, as new hair grows in the transplants, you've got hair that's growing almost the way nature intended it! Ex-television personality Hugh Downs is a great walking commercial for the grafts—although some recent bad publicity and a lawsuit resulted when one patient sued his doctor, claiming the doctor had told him the procedure was painless, but he thought it was just one big *ouch* after another!

FACING UP TO YOUR HAIR

I think women have very different opinions about mustaches and beards. I, for one, love a clean-shaven look on a really handsome man. I must admit, however, that for men who have a few flaws, a mustache or a beard can work camouflaging wonders.

The right beard can make a slight weak chin look very masculine. The right mustache can counter-effect a nose that's too long or less than perfect in shape. (On mustaches, remember two things: first, they tend to age a man. If you want to look younger, nix them. Second, always avoid a "droopy" look that can give your mouth a "down" line and your entire face a perpetually sour expression.)

I'm sure I don't have to add that you've got to stay superclean if you're going to sport facial hair. You have to be extra careful with your eating habits (and if you have both a mustache and a cold, you have to be extra careful to place your tissue over your mustache as you blow).

I hope you'll experiment with the look of your hair—

wherever you decide to grow it. A good stylist can be your best coach in improving the crowning glory of your packaging for the sexability market. Put yourself in his hands —look at his collection of style types—and trust his opinion.

WRITER'S PLOT WINS BY A NOSE

A brilliant writer I knew had millions of fans, but I suspected they were more admirers of the way he wrote than the way he looked! He had a nose that was hooked, was far too long, and appeared to be growing still. He had a rather long forehead, and he compounded the problem by combing his hair straight back.

He was surprised to find that he could get a "nose job" done by a competent plastic surgeon as an outpatient at the hospital. (His nose job consisted of removing the hump and shortening the nose.) Other, more complicated nose jobs, such as building a nose up (augmentation rhinoplasty) or reducing the size of the nostrils, might require up to four days in the hospital.

Our writer was somewhat surprised when the surgeon explained that the nose enlarges and elongates slightly with age. He feels he had the surgery done "just in time" and learned to comb his hair across his forehead to alleviate his "highbrow" appearance. His renovation was complete by the time his new book was published, and he said he had never enjoyed a public appearance tour as much as the one he undertook with his "new" nose!

MILLIONAIRE'S STOCK RISES
AFTER SAGGING LOSS

I knew a man who had about ten million dollars. It was enough to help him get along pretty well in the game of life, but all those years spent worrying about all that money must have left their toll. His kind-looking face had one of the worst cases of the sags I'd ever seen. His money helped him take off a few years, but certainly not all of them. We were good friends, and I finally told him I thought he could look half his age if he had all that excess baggage removed.

He seemed to feel better when I told him that women have their faces lifted at the first sign of a sag and that professional men do it all the time, too.

His plastic surgeon, after a preliminary examination, told my friend he was a perfect candidate for the surgery because "his tissue was very loose, very sagging. He had a sagging under his neck, in the jowl formation on each side of the face, exaggerated nasal labial folds, and bagginess of the eyelids." Doesn't that sound pretty?

The plastic surgeon explained that the operation would be done under local anesthetic, with a narcotic and a tranquilizer given to "render the patient amnestic" before surgery. Incisions are in different locations from those in a woman's face-lift, because of the difference in hair covering to hide the scars. The incision is made in front of and beneath the sideburns, across to the non-hair-bearing areas just in front of the ear, then downward and underneath the ear lobe, and again upward behind the ear to conceal the scar. (Scars are not usually noticed—unless the patient points them out!)

My friend learned that his hospital stay could be as short as two days, or as long as one week if he wanted to stay that long. He learned it was even possible to have the face-lift done on an outpatient basis, as long as the hospital had recovery facilities that permitted him to stay for three or four hours after each procedure.

He opted for a four-day stay, however. In a matter of several weeks after the operation—well, you've heard the expression "looking like a million dollars"? He looked every bit like the ten million dollars he was worth and seemed to carry himself with new pride and a new youthfulness, too!

BAGS ARE PACKED, BUT NO
ONE TO TRAVEL WITH

Pity the poor man whose "saddle bags" above his eyes are so bulging that he can barely see—or open his eyes wide enough so someone can gaze lovingly into them! I notice this a lot. I think it makes a man look like some sort of owl species, but it certainly does nothing to make him look like a lover!

Thanks to our adept plastic surgeons, however, those "saddle bags" above the eyes can be removed almost as easily as you can unsaddle a horse. Here's how a plastic surgeon explained the operation to me:

"Incisions are made in the upper lids, being carried out into the wrinkle lines lateral to the eye. On the lower lids, the incision is made 2 mm to 3 mm below the lower lash line and is also carried out laterally in the wrinkle lines. The redundant skin is removed. There is often bulging fatty tissue present in the lower and the lower upper lids. This fatty tissue is removed also, to prevent recurrence of bagginess. The amount of skin removed must be very accurately determined."

It takes only two to five days to transform an owly eyed man into a man who looks bright eyed, alert, and ready for love. The incisions are almost impossible to detect and aside from ridding a face of all that excess baggage, patients get other benefits as well. One patient told the plastic surgeon I interviewed that his visual proximity had increased at least 30 percent after the operation trimmed the bulges from above his eyes.

DOUBLE CHIN, DOUBLE YOUR
AGE

Pity the man who has no chin. He looks like a weakling, a Caspar Milquetoast. He can't turn his profile and get rave notices. He can't look aggressive as he strides into a room. He can't even buck himself out of a depression by responding to the advice, "Chin up!".

He doesn't have to grin and bear the situation—or cover the disaster area with a goatee! A receding chin is sometimes the result of heredity. Sometimes it's a condition that could have been corrected by orthodontic work. Today, it's a commonplace procedure for a plastic surgeon to turn a weak, receding chin into a firm, masculine one. It's the last thing she sees as her eyes travel down your face, you know, so your chin ought to leave her with a strong impression of your virility.

A plastic surgeon can create a manly chin for any male. The procedure is done either with the patient's own bone cartilage (in which case general anesthetic is used) or with silicone (which plastic surgeons also use to build up something a lot more sexy than your chin, in which case local anesthetic is used). The implant is inserted either between the teeth and lower lip or from the outside on the undersurface of the chin. The implantation can be done on an outpatient basis at the hospital, or it can require several days' stay if the bone cartilage for the implant is taken from the hip or the tibia.

DOWN WITH DUMBO

Big or protruding ears, alas, are not "the better to hear you with, my dear." Ears, like noses, tend to get larger and longer as we grow older. If you're now equipped with a pair that makes you look as though you could fly off with the nearest circus, your situation is so very, very easy to correct. With a simple operation, plastic surgeons can tack back protruding ears or decrease the size of long, drooping ear lobes. You'll wonder why you didn't have it done years ago!

Your plastic surgeon can work any number of minor miracles on you and your features. I do hope you've kept up with the times and have become more open-minded about what plastic surgery can do and who's currently reaping the benefits of plastic surgery.

The time has long since passed that only show biz types enlisted the plastic surgeon as their best friend in the youth-masculinity game. The image-minded business executive is now just as likely to keep his appearance full of youth and vigor through plastic surgery.

There is absolutely no reason to let age or a congenital condition diminish your sexability in any way.

You can get a list of fully qualified plastic surgeons by calling your local medical association.

You can shop among them until you're satisfied you have the surgeon for the job. You'll find that plastic surgeons never "sell" any procedure. They will take a lot of time explaining what they can do for you, and they'll have some before-and-after photographs of patients who have had surgery to correct the same problem you want corrected. The decision is up to you; don't you owe it to your sexability to investigate making the few changes that can make you a real physical winner?

THE DOCTOR WILL SEE
YOU NOW

You've been waiting for the routine physical examination you must have before you begin the exercise and diet regimen I've given you. Finally (after you've thumbed through all the office copies of well-worn magazines like *Today's Health*) you hear those welcome words, "The doctor will see you now."

You strip down to your briefs and socks and await your physical. Imagine your surprise when you discover "the doctor" is blonde, curvaceous, and very, very female! I hope she'll put you through the complete works to determine your physical condition, and that your bill of health comes out beautifully!

You're a little flushed as she coolly takes your medical

history. But if she were all-woman in addition to being all-professional, what would her handwriting reveal of her true feelings about your person? Let's take a peek at her notes:

"Baggy underthings . . . a little gray. He needs to change laundries or detergents or order eight fresh pairs of jockey shorts. Or maybe the ones he has on have been in his family for generations and generations?

"Ears, easy to find . . . haircut too short around ears. Needs to clip hair inside ears.

"Nostrils, ditto as per the hair . . . must prescribe plucking for this area.

"While checking knee reflexes, caught a decided aroma from area of feet . . . could he have worn the same socks two days running? Or hasn't he heard of foot deodorant?

"Had him remove socks and checked feet for firm arches, since he says he wants to start jogging. . . . Toenails misshapen . . . prescribed careful clipping to avoid a hospital stay for ingrown toenails . . . women don't like long fingernails or toenails on men. Needs to apply nail brush and file under toenails as well as fingernails. I pity his lover . . . if she's ever taken one of his big toes in her mouth to give it a little love nip.

"Notice he tossed his clothes on a chair instead of using the hangers in the examining room . . . bet he's a real slob to live with.

"Sent him next door for urine sample . . . he must have had to go badly, because after filling my little vial he made a vile amount of noise in toilet . . . he could have flushed toilet or let water run while urinating.

"Came on too strong when he saw me . . . same tired old jokes about my being a woman.

"General appearance healthy, but general lack of grooming apparent . . . hair overslicked with oil.

"Nice face . . . but can find no physical reason for him to breathe through mouth . . . gives patient a gawky expression that hinders appearance . . . ought to tell him this will eventually malform his mouth.

"Skin tone good, but overall dryness gives flaky appearance. Needs bath oil after showering.

"Quite hirsute chest and underarms . . . needs more powerful soap then men with little hair.

"Genitalia normal . . . but if he knew he was in for a total physical, how could he have missed a morning bath?

"ECG's normal . . . more exercise and better diet indicated . . . patient will return in two months for checkup after commencing diet and exercise program. Wrote prescription for Surbex-T and Theragran vitamin supplements. . . . Wish I could write one for his grooming as well."

THE PAUSE THAT RE-SEXES

Do you know that a lot of men shave or cut back the hair under their armpits? They do it for a very simple reason, too—to eliminate traps that hold perspiration odor. (And since underarms aren't where your "Samson" hairs grow, you're not in danger of lowering one tiny bit of your big supply of strength!)

Some men need to be doubly careful about their overproductive sweat glands. Some men need to use a deodorant *several* times a day, instead of just after their morning shower. Some men need, bluntly, to bathe more often, and to bathe with an antibacterial soap.

You owe it to her to bring everything to a halt a couple of times a day for the pauses that re-sexes. You can switch to a superpotent deodorant like Mitchum's. You can shower and wash thoroughly under the arms not only in the morning but after work as well. You'll never *think* of making love unless you've bathed very recently (you can always pull her into the tub with you, you know—"CONSERVE WATER . . . SHOWER TOGETHER." Nix too many hot showers, though, or your skin will get a bad case of "the flakes").

The same goes for maintaining a fresh, kissable mouth.

As with body cleanliness, this is a simple matter of a few minutes. A thorough brushing destroys mouth bacteria just as a thorough washing destroys underarm bacteria.

Your entire body will smell fresh and beautiful and lovable if you avoid spicy hot foods, foods with garlic, and coffee, liquor, and cigarettes. Such indulgences give you a body aroma that's anything but sweet!

Both pauses refresh you and give you better feelings toward yourself—and as far as she's concerned (no matter how eager she is to get you in bed with her), you've taken the time for the pause that re-sexes.

Your tongue can't be titillating unless it's tingly fresh to the touch! While you're busily attacking the bacteria in your mouth, don't forget to brush your tongue from front to back! All tongues have ridges and furrows, and your tongue may be more heavily ridged and furrowed than most. Bacteria can dwell in the tongue and cause odor— just as surely as they can proliferate elsewhere in your mouth. To be doubly sure, run your toothbrush over your tongue each time you brush. I can't imagine any loving couple kissing without using their tongues to increase the love play, can you?

RX: DEVOUR ONE FASHION MAGAZINE MONTHLY WITH ONE LARGE GRAIN OF SALT

I hope you'll flop down for an hour or so each month to catch up on the men's fashion news. You can take your pick from *Playboy, Penthouse, Esquire,* and *Gentleman's Quarterly.*

Look at the pretty pictures and study the pretty prices. Make a special note of where the clothes you like are available in your city. Then do a little mental juggling and pretend you are the fashion editor of the men's magazine you're reading. You're struggling against a deadline, and you've got to come up with something that will make the competition look as though they were still promoting knickers and Buster Browns.

What do you do?

You manufacture an attention-grabber, that's what. You call a couple of your favorite mod manufacturers and get them to dress your male model. He appears in full glowing color under a headline screaming, "Looking ahead to Spring in Palm Springs," and he's wearing a lace see-through shirt in subtle mauve, bright yellow-checked hopsack trousers, a four-inch white patent belt with an enormous gold buckle, matched by a two-inch white patent watchband supporting a revival of the Dick Tracy timepiece, and black and white pointed-toe shoes which you, as editor, predict will make a "smashing comeback for spring."

As an editor, you've done it! You're certainly going to be talked about this month! In all honesty to you, let's say that your information was, to some extent, quite correct. Maybe one of your five wardrobe items might, in a certain season or a certain year, be just fine for Palm Springs. But what about the faithful reader in Des Moines who takes all your "fashion pictures" at face value and puts himself together in faithful reproduction of your photographs for his once-in-a-lifetime trip to Palm Springs? He'd be a walking disaster.

Before you get carried away with any new trends a fashion magazine reports, you owe it to your body, your buddies, and your budget to consider who you are, how you live, and just where you're going. What might be great for Disneyland East in April may make you look like a refugee from a Disney cartoon in Des Moines next summer.

I'm not saying you should be ultraconservative, but I do want you to dress in good taste. Have fun with a few purchases, but for the bulk of your clothing expenditures, put a lot of thought into how often you can wear them and how well suited they'll be for your daily life-style.

No fashion editor would ever earn his salary if his magazine or newspaper came out month after month with the statement, true as it may be, "The gray flannel suit is still tasteful and suitable for your every need." You'd quit reading him in a minute.

Read the fashion news. Get excited about new trends. Make lists of items you'd like to acquire.

But for sexability's sake, *NEVER* put it all together at once. Take that grain of salt. If you choose a mod, loud shirt, keep your tie and jacket on the quiet side. If you're sure a mauve lace see-through shirt is what's really in for a resort area, couple it with white trousers and white shoes. Take just one or two of your new ideas at a time—don't be like a woman who has to wear every piece of jewelry and every fur in her closet just to show off how much she owns!

THE PROS IN YOUR LIFE:
DOCTOR, LAWYER, CLERGYMAN—
AND CLOTHING SALESMAN

We're living in an age of specialization—we need a couple of types of doctors and dentists; we need corporation lawyers and divorce lawyers; we need a clergyman and a tax expert.

I want to nominate one specialist whose services will cost you absolutely nothing. He's the man who can advise you on your wardrobe purchases so you can create the package that sells you as a person!

He's the fellow who keeps up with the latest trends—and I don't mean the latest fads. He's the man who has the nerve to lose a sale temporarily and say, "Mr. Customer, this suit isn't for you." He'll steer you toward the right accessories—even if you bring in a favorite pair of basic business suits from two years back and say, "Mark, I need some help to select the best new shirts, ties, and belts to go with them."

This marvel of a man will call you when a new shipment arrives with a couple of suits he thinks were made for your way of life. (Yes, and he'll even have the nerve to tell you to hold off on an important purchase because it will be marked down considerably in three days.)

Like any professional in any profession, this salesman will do all these favors for you because he's smart enough to know he can win your confidence so that you gradually begin to trust him for all your clothing needs.

You can go the other route, of course. You can amble into a dozen different stores and shuffle around until some overdressed young fellow pays some attention to you and pushes off on you (and your charge account) the things the store is loaded with and trying to dispose of before they're totally dead. A lot of salesmen advised you to stock up on several Nehru jackets with a turtleneck shirt and pendant "necklace" for cocktail wear rather than just one for fun. The suit is now seeing its second or third go-round at Goodwill Industries, and the salesman is probably on his fifth or sixth—or tenth—new job since he sold you the ensemble.

FINDING YOUR PRO

Just how do you find such a PRO in men's fashion, one who won't ever, ever let you down?

You do the same thing you do when you want to find a great restaurant, with a marvelous maître d' who guides you in your choices. Only insecure men never say, "I need some help."

Smart men, no matter what their income, first search out the *best store* in their area. Then they go in and announce their intentions. Neiman-Marcus, the ultrahigh fashion store which started in Dallas and now has branches spreading all over the country, takes justifiable pride in fulfilling orders by letter. I'll quote one from a young woman: "I have only three hundred dollars to spend on my wedding gown and trousseau. Please help me." Voilà! The entire Dallas store rallied around the young bride-to-be. They met all her needs—including her slender budget —and made a customer for life.

All good stores which place great value in customer service would make the same effort. Seek one out. If your bank balance is slim, don't amble in and pretend you have thousands to spend. Say, instead, "I've got —— to spend, and I'm starting at a new job that requires that I have the best basic wardrobe I can. What would you suggest?"

If you should end up with a salesman who seems less than efficient, ask to see the buyer in charge or the assistant buyer. You'll be amazed at how quickly you have talented retail experts at your disposal, all trying to make you the best-dressed fashion picture you've ever seen. It's a challenge to them . . . and they get a fringe benefit besides:

1 They're getting the personal pleasure of building your confidence and image by suggesting the proper clothes.

2 They know that as your image grows, so will your income.

3 They also know that you'll continue to patronize their store as you continue to improve the wardrobe packaging that makes you the man she'll look at again, and again, and again.

SUITED TO YOUR NEEDS— AND YOUR BUDGET, TOO!

Let's pretend that you have spent two muscle-building years as a longshoreman, and suddenly were offered a job as a Madison Avenue junior executive. You'll have to toss overboard your work denims and build a wardrobe suited to your new job.

What would be your basic wardrobe?

Suppose you could immediately afford only one suit. What would it be?

Interviewing many outstanding retailers gave me one answer:

133

THE BASIC SUIT, THE ONE THAT WILL GIVE YOU A VARIETY OF LOOKS WITH DIFFERENT ACCESSORIES, IS THE GRAY FLANNEL SUIT.

My fashion experts said to start with the gray flannel suit and to add a navy blazer for casual wear. Beware, they said, of current promotional items (such as patterned double knits). You won't get the versatility from such novelty items—and they may be a passing fancy of fashion. You can indulge in such items only after you build a wardrobe that will wear well for all your basic needs.

Listen to what the experts have to say about "your" gray flannel suits:

"Flannel is going to be even more important than it once was," said Neiman-Marcus's Neil Fox. "Flannel used to be more basic, but I think this year it is more meaningful from a fashion point of view. It will probably be the single most important woven fabric in the market.

"A man could take a gray flannel as his basic suit and go to navy blue for the sport coat. The navy blazer could be a double knit, or it could be a dacron-wool blend. For the fall of the year, if a man can afford to have more than one, he would have a flannel-type blazer or a worsted hopsack.

"I wouldn't advise a man who could afford only one suit in his wardrobe to go to navy blue. Most navy blue fabrics, unless the fabric is flannel, tend to develop a sheen after a while. The navy blue double knits that are on the market tend to have a purple hue, and not be a true blue."

Neil Fox admitted that gray didn't work too well with some people's skin tones, but he said, "A man can offset that with his furnishings and accessories. Gray is practically a neutral color, and you can go practically any direction in accessorizing gray. You can certainly go to yellow with gray—or you could do the things people expect you to do with gray. If you want to dress gray up at night you can go with the white shirt and a small-checked tie. That's very dressy. You can go into pink tones with gray, too."

Saks's Sid Mayer also nominated the gray flannel suit and

the navy blue blazer as his number one "starter suit" choice. "It should be single breasted," he told me, "and the blazer should be navy blue, either single breasted or double breasted."

What the man wears with the blazer, Mr. Mayer told me, is a matter of his individuality and his taste. "If a fellow has a certain taste level, he would wear the gray flannel suit with the blazer—or he would wear a plaid pair of trousers or a fancy pair. I personally like something in the tan family, in a cavalry twill or a hopsack, or in a camelhair look. Any of these colors would be beautiful with the navy blue."

Sid Mayer reported that the blazer, in many big businesses, is considered proper workday attire. "Saks Fifth Avenue now allows employees to wear blazers to work," he said. "I would say that a pair of gray flannel trousers and a navy blazer—if the man's firm has no objection to it —is fine for business. If a man works for any agency, or in any of the theatrical areas, even in medicine . . . this is a fitting work outfit."

You'll build from this simple nucleus, of course. You'll experiment with various accessories and various colors to change your basic looks. You'll supplement your basic gray as you can afford to do so.

But start with something that gives you a good point from which to build. Take a tip from the woman. If she has to have just one cocktail or after-five or theatre dress, it will still be a little "basic black" (with pearls!). Only when she's got all the basics she needs for work, play, and evenings out, only after these basics are assembled, does she happily veer off into mad patterns and colors—or allow herself a fling at the latest fashion flash.

Starting with the basic suit—the costume suited to your everyday needs—is like buying taste insurance. You may not win rave notices for your enormous array of clothing, but at least you'll be dressed in good taste wherever you go.

TYING IT ALL TOGETHER

I just know there's a special place in heaven for the man who invented the necktie. What was his name? Quinby Cravat? I don't know, but I do know that we owe him a big debt of fashion gratitude. How else could your gray flannel suit go to a funeral in the morning (with a blue knit necktie suggesting an appropriately somber air?), to a client luncheon at noon (wearing a wide tie in a happy yellow to match your socks, a yellow that says, "I'm cheerful because it's Monday, and I'm going to show you how we can make a lot of money"), and to an alumni cocktail party at five (wearing a tie in a shepherd's check —very dressy)?

Your collection of neckties can make your wardrobe.

If you're smart about selecting them, and about coordinating them with your shirts, you can actually give a lot of new and different looks to just a few basic suits.

If that isn't enough, ties can enhance your complexion. An accent of color that close to your face can make your skin tones look fresher (or deader). The proper color can spark up your suit or tone it down. Women play all sorts of tricks with scarves and basic dresses to stretch their wardrobe dollars, and you should learn to be just as cagey with your selection of ties.

THE TIE THAT BINDS
(YOUR WARDROBE)

Here are some "how-tos" that well-known retailers gave me about ties:

1 Save the bold-patterned tie to use as your fashion exclamation point all by itself. Don't pair it with a bold-patterned shirt or suit.

2 The reason you're suddenly seeing so many

136

different *lengths* in ties is not that the designers have started creating for men of various heights. Rather, it's because the varying *widths* in ties make the ties thicker and the knots wider. If ties were the same lengths as they were in the past, the tie would end up very high on the torso, instead of extending to a more graceful level.

3   With both wider ties and thicker fabrics being used for your ties, be sure to buy ties that are long enough—or you'll look as though someone took a pair of scissors to your tie. If you want the tie long, of course, you start tying with the short back end—and if you want the tie shorter, you start tying with the longer end.

4   The narrow tie is out right now, but you can guess it will be back as the pendulum of fashion makes its swing to reverse current trends. Don't get too hung up on the game of wide-versus-narrow. Look at your total fashion picture. Neiman-Marcus Vice-President Neil Fox (in charge of men's wear and a superb dresser himself) admitted to me that he didn't exactly get physically ill at the sight of a narrow tie. "Men have to remember the balance and proportion of what they're wearing," he said. "If a man is wearing a relatively narrow lapel suit, and is still wearing narrow ties, at least he's still in balance. Of course, if he's gone out and bought himself a new suit, with at least a four-inch lapel, and he's wearing a three-inch necktie, he'll still be very out of balance fashionwise."

PANTING FOR ATTENTION

Most women still want you to "wear the pants in the family," and they want those pants to fit well enough to give at least a hint of your male physique.

I talked to a lot of outstanding men's fashion retailers and got their clues as to how you should pick and choose

if you want her panting after your body. (And none of
this implies that you should be running around like a
ballet dancer, as beautifully developed as they are, with a
tight-fitting leotard revealing every bit of your male
physique!)

Here's how the fashion picture is going to shape up between your ankles and your waistline:

1 "Trouser styles vary almost from one six-month
   period to the next. These are minor variations,
   like whether the pants bell a little bit, or flair a
   little bit, or peg a little bit. A man won't be an
   outcast if he doesn't keep up with these minor
   variations, but it's sort of like visiting your dentist.
   If a man misses too many appointments, he's in
   trouble. If a man lets too many changes in cut
   pass him by, he will look dated."

2 "Men should definitely consider their body
   structure before buying trousers or suits. A heavy
   man should avoid the thick, nubby tweeds; but,
   on the other hand, a very thin man would add
   some dimension to his figure by buying pants and
   trousers in heavier fabrics."

3 "One of the latest looks in trousers is for the front
   of the trouser to be touching the top of the shoe,
   with the back of the trouser extending down the
   heel for some distance. There might be a difference of three-fourths of an inch from front to back.
   But the front look is crisp—just touching the shoe
   top."

4 "Flared trousers don't necessarily have to actually
   flare at the bottom. Actually, this is an optical
   illusion; if a trouser leg is the same width at the
   knee and the bottom, when a man puts it on, the
   trouser will still look flared."

5 "The stovepipe trouser look is out. The flare gives
   a lot of accent. Pleated trousers would be considered high fashion, whereas before they were
   considered the most conservative."

6 "One of the worst criticisms I have is that I see
   so many men—more in America than in Europe—
   wearing their trousers overly short. Europeans

somehow seem to have a better sense of balance and symmetry about their trouser lengths. The look of today is to keep the trousers long enough so that there is no break, with the trousers longer in the back than in the front."

7   "A man may be basically 'safer' with a plain trouser bottom. However, for the fashion extremists, and the Left Bank types, there's still a mood and a feeling for cuffs. If a lot of lapel is showing—the sort of look that some of the younger men in New York are wearing right now—the two-and-one-half-inch cuff is the right look. It's not a look for a short man, of course, or a man with short legs."

8   "The most important thing is that Americans get away from the same feeling about trousers that they once had about jackets. Any man is going to look ten or fifteen years older, and that many years behind the times, if he wears trousers that are baggy. Any man with a nice physique wears fitted trousers, just as any woman who isn't ashamed of her figure wears a well-fitted dress."

GETTING YOUR FOOT IN HER DOOR

Your best foot forward can be the one she loves to look at—providing it doesn't remind her too much of her grandfather's!

The day has passed in which the lace-up shoe was the only proper one for business.

Today, if you want a pair of shoes to wear for both day and evening, the shoes will be slip-ons.

"Americans once felt they had to have a big, heavy, serviceable type of shoe, but that's changing now. Of course, some men are overreacting to the trend and have gone to the garish and the overdone," said one leading men's fashion figure.

"I stopped wearing Gucci loafers four years ago. . . . I

think the feeling for brass on the foot is over from a fashion standpoint. Look now for the covered buckle, or just plain tassels, which I like. This is not to say the Gucci look is out, but it must be the real McCoy, not the fifteen-dollar knockdown."

My expert said he could sit in a hotel lobby anywhere in Europe and tell which men were Americans—without even raising his eyes from the floor!

"A few years ago, I could look at those trousers up around the ankles somewhere, and see those big, overpowering shoes, and just know I was looking at an American."

Try for a relaxed look in your footwear; she doesn't like the clunky look on your feet—any more than you like the clunky shoewear she bought a season or two ago. Listen to what the experts say, and realize that the Gucci moccasin is going to be the classic of the 1970s, just as the penny loafer was of decades past.

You will vow to give your shoes a daily shine, to let each pair "rest" between wearings, and to insert shoe trees immediately on taking your shoes off. You'll prolong their life quite a bit if you do this—and save a lot of wardrobe dollars.

SOCK IT TO THEM

Short socks are out.

Finished.

Through.

And so are you, if you continue to wear them.

Helen Gurley Brown told me that if she finds herself sitting next to a man wearing short socks, she resigns herself to the fact that she's going to have absolutely nothing to say to him!

Your short socks belong somewhere with pictures of you in your crew cut, your button-down shirts, and your college letter sweaters.

140

The look of today is no leg showing (unless you're totally sockless in a resort area—more about this in a minute). The sock reaches well over the calf, insuring that the line of the trousers-to-socks-to-shoes will be unbroken.

I've given long socks to my dentist, to business associates, to doctors, to my minister, and to a lot of other men I care about who had been ruining their looks with short socks. If you don't believe how bad it can be, watch some television talk show, with all the male participants sitting around with their legs crossed. If you can find one showing an expanse of leg, you'll see what I mean. I'd rather see a man barefoot!

One word about the longer socks. Ask that they not be dried in a dryer—or, if your housemistress insists on drying them this way, buy her some antistatic fabric softener or look for "static-free" fabrics in your long socks, because if the trousers you're wearing are partly synthetic, you will feel your trouser legs crawling up toward your knees—hardly the look we're striving for.

Is there a place for all your short socks?

Decidedly yes. If they're white, I could stand to see them on a tennis court. If they're dark, they'll make dandy dustrags!

BAREFOOT IN THE PARK—
AND OTHER FUN PLACES

Bare feet are sensuous.

And loafered or sneakered feet lose all their appeal when you clobber the look with socks!

You wouldn't, I hope, wear lace-up leather shoes with a pair of Bermuda shorts, would you? The look would be horrible because it would be like going on a two-week cruise and working on your law briefs the entire time!

When you're out in a resort area, LOOK DRESSED FOR FUN. Make your footwear casual—loafers or sneakers. And KEEP YOUR FEET SOCKLESS, even after dark.

141

It's the in thing to do. (Be sure your feet and legs are tan, or use a tanning cover-up to create the resort look that's perfect.)

But stow your socks; save them for when you return to your workaday life.

TANNING YOUR HIDE

Many fashion looks for men are marvelous because they expose the man's virility for all to see. Think of the sockless look with highly polished loafers for resort areas . . . think of sport shirts with the top button left undone, exposing a manly chest . . . or the look of muscular arms on the tennis court.

There's only one qualification for all these areas to be showing and still be sexy. THEY'VE GOT TO BE TAN. NOTHING, but NOTHING, turns a woman off as much as an expanse of lily-white skin on a man. That's supposed to be her domain, not yours!

I'm not advocating torturing your skin and aging it prematurely by overexposing it to the rays of the sun. If you live in one of our more tropical climes, you can even induce skin cancer by overdoing the sun scene.

But you can give your arms, legs, face, and body a healthy-looking glow. If you can't tolerate the sun, you can certainly use one of the commercial preparations such as Man-Tan or Q-T. They'll do a lot to transform those fishbelly white calves into the legs she loves to look at. A moderate amount of sun is good for everyone, and many of the commercial preparations provide the protection you'll need to reap that beneficial vitamin D without harming your skin. PABA ointment has been found especially effective to protect the skin of blonds and redheads, and there are effective lotions that screen ALL the sun's damaging rays from your skin. (Ask your druggist or dermatologist about these, especially if you're the sort who blisters instead of tans.)

Whatever route you take, try to give the glow of sunlight to that body of yours. You'll look like one who loves the *outdoor* way of life . . . and when you look this way, you're much more likely to turn her on for an *indoor* way of life that's lots and lots of fun.

UNDERNEATH IT ALL

What goes underneath your shirt?

Nothing—if you're aiming for greater sensuality.

Men with the most sexability told me they wore their last undershirt sometime during seventh grade—when they got big enough to stand up to their mothers and refuse to put one on even if it meant catching cold.

Women don't like to see the lines of an undershirt showing through your shirt. They like to feel that your bare chest is very, very close. Fashion experts say that wearing an undershirt has nothing whatsoever to do with making your shirt look better or wear better. "We still sell them, of course," one retailer told me. "We're in the clothing business, after all. But I don't think undershirts are attractive or necessary in today's world. I think it started with Clark Gable in *It Happened One Night*. That was sort of the beginning of the end for the undershirt."

I'd like to add that undershirts make your waistline look thicker and that they can ruin the lines of a fitted, tapered shirt.

THE LONG AND THE SHORT OF IT

As for undershorts, I found no distinct trend. One man said he still likes the boxers, short and fitted, and one size smaller than his trouser size. He said he wears white, beige, or light blue, and "that's as far as my underwear adventures have gone." Other men favor traditional

143

jockey shorts, and several I talked to liked the knit "bi-kini" shorts.

Almost all my interviewees (after going into elaborate detail about what sort of pajamas they liked best) admitted that they'd much rather sleep in the nude.

"I buy my wife all these sexy gowns, and she's bought me pajamas that have made me look like everything from a Chinese emperor to a Russian hussar," one man laughed. "And she sleeps in the raw—and I sleep in the raw. In case of a fire—if we had time to find the pajamas and night-gowns, I guess we'd look well dressed for the firemen's benefit."

Whatever your choice of underthings, I can't say enough about the freshness of the clothes you wear right next to your body. Well-dressed men always have their shorts pressed for an extra well-dressed feeling. When you pay two hundred dollars for a suit, why do you begrudge spending a dollar on a new pair of undershorts? You won't feel like a well-dressed man until you know you're well dressed from the skin out!

LET ME TAKE YOU TO
THE CLEANERS

Even if you have no woman to sew on your buttons and press your clothes, I want you to have crisp, neat trouser creases and impeccably pressed jackets. The absence of wrinkles goes a long way toward making your fashion image.

If you aren't going to maintain your clothes, due to so many demands on your time, I'd rather see you in perma-nent-press Levis and permanent-press T-shirts. What a shame that you spend a week's salary on a suit and then let its rumpled look betray your fashion image. You might as well rumple up five fifty-dollar bills!

Your own closet and your neighborhood cleaners can be the best friends your wardrobe ever had. Let's start with your closet.

"I'd rather see clothes cleaned and pressed a little less, and better cared for after each wearing," one retailer told me. "Cleaning and pressing is necessary, of course, but it does decrease the life of any garment if it's overdone."

He told me that every man should collect hangers that conform to the lines of his suit jackets. "The hangers should be wooden, and the shape must be right for the jacket," he emphasized. "You can't hang a jacket on a straight wire hanger and expect the suit to stay in good shape." Hanging up trousers and jackets immediately after wearing adds a lot to their probable recovery from the day's wear. Never overpack your closet, or you're asking for more wrinkles during what should be the "resting period" for your clothes.

Every man should be on intimate terms with his neighborhood dry cleaner. (Especially if she happens to be a shapely blonde!) In any event, he should know what services his dry cleaner can perform to keep his clothes in A-1 condition. He should make it clear to the cleaner that, after all, he's a man who cares about his appearance, and that since he's going to be tossing quite a bit of business in his direction, he expects the following services to be performed automatically:

1 Dangling or just-beginning-to-loosen buttons firmly reattached. (That old adage that a stitch in time saves nine is only half true. It also saves many the loss of a button, a minor irritation which can occur at the most inconvenient times.)

2 Restitching of any seams or pocket flaps or cuffs that are beginning to pull loose. If you wait until they're noticeable, how many other people have also noticed, and how do you think your fashion image has fared in the meantime?

3 Speedy alterations as you continue your path toward greater fitness with my exercise program. Never wait until pants become terribly baggy or suit jackets hang like those on impoverished scarecrows. A good cleaning establishment has alteration experts on hand—and since you're such a good customer, it won't keep your suits tied up

for a couple of weeks while the minor alterations
are performed.

When I said that such services should be performed
"automatically," you'll notice I didn't say "gratuitously."
Of course, you're going to pay for any such extras. But
you're going to pay for these extras willingly because they
protect the considerable investment you have in your
clothes. (If you want to consider the alternative, that of
maintaining a full-time seamstress in your own home, just
remember a w-i-f-e is going to cost you quite a bit more
than a few tiny alterations bills!)

One thing you can do at home or on out-of-town trips,
however, to keep your clothes looking fresh from the
cleaner's in between trips is to use one of the new steam
"pressers." They retail for under twenty dollars, and several clothing experts told me they are much easier on your
suits than conventional pressing. You simply fill the
"presser" with tap water, hang the garment up in the bathroom, run the presser over the garment, and watch the
wrinkles disappear. It's the new wrinkle in clothes care,
and I think every man—bachelor or with live-in wife or
(wardrobe) mistress—ought to invest in one.

THE BEST AND THE WORST
OF THE DRESSERS

I'll end this section of dressing and grooming with a poll
of my experts. Here's how they voted men in various professions as far as their dress. If your profession isn't listed,
maybe you and the other men in it haven't made any
impression at all—now wouldn't that be a shame? Here
goes:

THE BEST-GROOMED LIST

> *Actors.* (Especially when they don't overdo the show
> biz bit with their dress.) George Hamilton's a good
> example of a fellow who knows how to put everything
> together without overdoing it.

146

Upper-Echelon Military Men. All that training doesn't go to waste or to waist. Women have a thing for the wonderful bearing and crisp grooming that rich, retired admirals and generals have.

Advertising Execs. Not necessarily the mad creative souls in the art department, but the agency heads and account executives. They keep up with fashion and know the importance of image projecting.

Some Lawyers. The more courtroom drama they produce, the better the packaging.

Clothes Designers. They know a good line when they see one.

Homosexuals. They seem to knock themselves out more to stay on the sexability market than their heterosexual friends—how strange!

A Few Doctors. Like Marcus Welby, M.D.—he's divinely groomed and dressed. Most of the rest are altogether too rumpled, especially when you consider the incomes they have and could devote toward shaping up their grooming.

Athletes. How their image has changed in the past few years! I think they're suddenly among the snappiest dressers in America, and they have the physiques to make their clothes look great on them.

Young Corporation Executives. A lot of the older ones still look to me like something out of a Dickens novel, but the ones smart enough to have made it to the top before they hit forty are usually smart enough to dress the role of a wheeler-dealer.

Interior Decorators. The successful ones couldn't have all that sense of style and color in one area and ignore it in another—their own dress and grooming.

Golf Pros. Swinging. They've got a country club milieu to swing in, and they usually dress the part.

Men of Various Occupations Living in New York or Los Angeles. They're where the fashion action starts, so they catch on to new trends first.

THE WORST-GROOMED LIST

Men of Various Occupations Living in the Midwest. Despite television, despite national news services, these fellows hardly ever pick up a trend until it has

been replaced by a new one. You know the kind. Just
as he lets his crew cut grow out and buys his wide-
lapel shirt, you can bet your fashion dollar that the
short look in hair and the crew cut will both make a
dramatic reappearance. How sad!

Newspapermen. They always look covered with
carbon, and their trousers always look baggy. How
can they be so bright and dress so dully? Maybe
they're so busy shaping our thinking for us that they
haven't time to shop!

Priests, Ministers, and Rabbis. I don't think the Bible
ever said, "Thou shalt not look handsome and
well dressed." I think these fine men can certainly maintain the respect of their congregations
and still be "holier than thou" without having to
look "seedier than thou."

*Almost Every College Professor and High School
Teacher.* Yes, even those living in my trend-setting
cities. Did they have to take a vow of sloppiness
before they could get their tenure? I'd sentence every
one of them to a sabbatical for study of this book!

In the middle are a lot of wonderful men in wonderful
professions. There are body builders and local businessmen,
small-town mayors and big-town congressmen. There are
college students (who can either look like shaggy animals
or fashion plates, depending on their current state of rebel-
lion against the fashion and grooming establishment, and
whether that honey-haired girl in psychology smiled at
them during class). There are hairdressers and stockbrokers
and dentists and butchers and you and you and YOU.

And for each man, there's a woman wishing you'd change
a few things to make you the man she's going to want to
be with—forever. You can do it yourself, you know, and if
you do it before she turns toward a man who's making his
package more attractive, you'll make three people very, very
happy. You'll be happier. She'll be happier. I'LL BE HAP-
PIER, too, because I'll know that your improved packaging
is helping you win at the game of love.

GRADE YOUR SEX APPEAL

Are You Sexy?

		Yes	No
1	Do you occasionally wear your shirt unbuttoned halfway down the front?	( )	( )
2	Do you sleep nude?	( )	( )
3	Are you a "first draft choice" on the measurement chart?	( )	( )
4	Do you have a good posture, with your pelvis pushed slightly forward and your hips tucked under with the chest pushed up and away from the waistline?	( )	( )
5	Can you effectively use eye language?	( )	( )
6	Do you have a sensuous voice?	( )	( )
7	Do you wear loafer-style shoes?	( )	( )
8	Do you use a body stance to provoke sexual desire?	( )	( )
9	Do you have an attractive hairstyle?	( )	( )
10	Are your clothes always freshly pressed and your shoes shined?	( )	( )
11	Do you have a fresh, tasty breath?	( )	( )
12	Do you use a lip ice to keep your lips soft and desirable?	( )	( )
13	Do you use cologne?	( )	( )
14	Do you have good, vibrant health?	( )	( )
15	Do you wear well-fitting undershorts?	( )	( )
16	Do you wear long socks for dress?	( )	( )
17	Do you always have your shirt collar and cuffs showing with a suit coat?	( )	( )
18	Do you feel secure?	( )	( )
19	Do you avoid overimbibing?	( )	( )
20	Do you generate warmth?	( )	( )
21	Can you keep your cool?	( )	( )
22	Are you attentive?	( )	( )
23	Can you carry on a conversation for thirty minutes without once saying "I"?	( )	( )
24	Are your teeth free from stains?	( )	( )

149

		Yes	No
25	Do you wear contacts or tinted glasses (if glasses are needed)?	( )	( )
26	Is the hair in your nose and ears trimmed?	( )	( )
27	Do you brush your tongue every day to remove bacteria?	( )	( )
28	Do you pay attention to your pedicure?	( )	( )
29	Do you keep unruly eyebrows trimmed?	( )	( )
30	Does your body have a pleasing aroma? Would you say it tastes good?	( )	( )
31	Do you shave before making love?	( )	( )

Do You Have Unsexy Habits?

		Yes	No
1	Do you talk loudly?	( )	( )
2	Do you wear what you want to whether it's appropriate or not?	( )	( )
3	Do you touch your teeth or pick your teeth in public?	( )	( )
4	Do you stretch in public?	( )	( )
5	Do you spit in public?	( )	( )
6	Are you a "clothesline sitter" with both arms outstretched?	( )	( )
7	Is your car or apartment untidy?	( )	( )
8	Do you brag?	( )	( )
9	Do you have smelly feet?	( )	( )
10	Are you too skinny?	( )	( )
11	Are you too fat?	( )	( )
12	Are you flabby?	( )	( )
13	Do you wear clothes that don't fit?		
14	Is your posture bad?	( )	( )
15	Do you overindulge in coffee, booze, cigarettes?	( )	( )
16	Do you smoke cigars?	( )	( )
17	Do you ever have bad breath?	( )	( )
18	Do you make love without having bathed?	( )	( )
19	Are you an incessant talker, concentrating on me, my, mine?	( )	( )

		Yes	No
20	Do you talk about past dates and conquests?	( )	( )
21	Do you carry a chip on your shoulder?	( )	( )
22	Are you often irritable or impatient?	( )	( )
23	Do you put out your cigarette butts in coffee cups or dinner plates?	( )	( )
24	Do you light a cigarette between courses?	( )	( )
25	Are you fidgety; do you pull your ear lobe, tweak your nose?	( )	( )
26	Can the person sitting next to you ever hear you eat?	( )	( )

I made my answers pretty easy to guess, didn't I? I hope you'll build up your sex appeal until you can honestly come up with twenty-six Yes answers on the "sexy" questionnaire and twenty-one No answers on the "unsexy" questionnaire.

Some questions are big ones, and some may seem relatively unimportant. But they all add up to your sexability quotient—the sum total of how your every action turns her sexual feelings toward you on or off.

You should have 100 percent Yes answers for the "Are You Sexy?" quiz. You should have 100 percent No answers for the "Do You Have Unsexy Habits?" quiz. Anything less than 100 percent on either test, and your sexability rating is suffering. You've got to build up your sex appeal and start doing it right now. Until you do, you'll never fly high with Ginnie. You'll never fly high with Marlene. And you'll never even get your sex machine off the ground until you make a major overhaul. No fair reading the next chapter until you do!

Chapter 6
Sex–The Sport Every Man Can Excel In

Not every man can become a first-rate athlete, but every man, if he is willing to train hard, can learn to excel in the most popular sport of all—SEX.

The rules are going to be a bit confusing to those of you who are regular fans at the football or baseball stadium, simply because you've become accustomed to rooting for your team to score quickly and finish first.

And when you've been rooting for your team to score first, it might be difficult for you to switch attitudes and root for the other team, her team, to score and hope for a tie game. That's the way it should be in the great game of sex.

You become the winner in the great game of love when you say, "Baby, I want to help you score two touchdowns before I even have the ball." Once you learn this, and learn *how* to insure her success, you've taken a giant step toward the Hall of Fame!

As natural as lovemaking might seem, you are not proficient at it by instinct alone, just as you're not born with the instinct to become a champion in the sporting world. In both instances you must be skillfully developed into a good athlete. And no one can train you successfully unless you have the desire to be good—to be extraordinarily good!

To get "peak enjoyment from sex, you must give peak performance!" No athlete would consider approaching a contest without working to be in perfect physical condition, without learning all the rules in the book, without learning what to expect from his coplayers, without mastering all the techniques and planning all the strategies that will allow him to play at his very best.

So, how do you play the game of sex?

It would take a lifetime of loving to "happen upon" all the wonderful techniques that make a woman's body respond to yours.

If you plan to "let nature take its course" until you become a good lover, you're right in the class with the caveman who did without cooked meat for eons and eons until a flash fire accidentally cooked it for him! Maybe the caveman never knew what he was missing, but the culinary art is one thing and the art of sex is another. And you haven't GOT THAT LONG TO LOVE! If you care about becoming an excellent lover, why leave anything to chance?

Every major league athlete gets to watch a rerun of all the plays of the game so he can correct his mistakes in the next game.

If you could see all your past mistakes, and had a helpful coach at your side, it would help you quite a bit, wouldn't it? But since (I assume) no movies have been made of your past games, we're going to start from scratch —and one of the best ways I know to do this is to let you learn from hearing what the professionals say who play the game so well!

My training camp will give you the expert advice of some of the best coaches in the sport. Listen to them, and you can make the Hall of Fame!

Remember that I asked you to stay in my training camp for one season? To work out and put yourself in top condition, and eat a diet that will give you fuel you need for power and stamina in the game?

I warned you, too, not to feel overconfident just because

no woman has ever told you that you belong in the Little League in the sport of sex. There is only one person who can critique your lovemaking, and that person is your partner in sex. *She* usually suffers in silence, for fear of bruising that tender ego of yours, and she hopes that somehow you'll improve the game without her having to tell you!

SEX, THE "SECRET SPORT"

You can decorate your stadium, your playing field in any way you wish, but be sure you know the rules of the game!

If you played baseball the same way you play sex, the term home run would soon vanish from your vocabulary. Abner Doubleday (the inventor of baseball) set up the game so well that generations of tough umpires have made sure everyone understands the rules of the sport.

But until a few Kinsey-enlightened years ago, only a real undercover agent could have written out the playing techniques in the great game of sex.

What a situation!

Here's everyone's favorite participation sport (would-be lovers outnumber baseball players at least 100,000 to one), yet the rules and techniques of sex are still so hush-hush that at least 95 percent of the players are still in the minor leagues.

Suppose we played baseball without rules. Imagine yourself as a rookie, confronting your first major league coach!

> ROOKIE: My farm club coach told me I had great potential. I'm really hoping to make it big here in the majors.
> COACH: Yeah, I'm sure you will, kid.
> ROOKIE: Of course, there are many things I've still got to learn . . . how to warm up right, get in shape for the game, and get along with my teammates . . . all the things I will need to be successful.

COACH: Look, Sonny, it's just a matter of doin' what comes naturally. Besides, in this league, over three-fourths of your games will be played in the dark, so if you make any errors, they will be hard to spot.

ROOKIE: If you say so, Coach. But what if I don't perform? What if I misjudge and swing too early and miss the ball? What if I can't get completely warmed up before game time? What if I run out of energy and just can't make it to the final innings?

COACH: Look, kid, do me a favor and just knock it off? You start worrying too much, and you'll just tighten up. Relax—let nature take its course! Get it?

ROOKIE: Aw, Coach, you mean you aren't going to teach me *anything*? I don't want to overtrain, but you *gotta* help me because I sure as *hell* don't want to strike out my first crack at bat.

COACH: Listen, Sonny, I'm *damn* sick of players like you. It's your attitude. What have you got, a dirty mind or something? Didn't your dad ever tell you that gentlemen never talk about things like bat, ball, scoring, etc? I got pretty good reports on you from Crunch City. But they never told me you were a dirty-minded type that had to talk about strategy. Who taught you the facts of baseball? Listen and listen good: IT JUST ISN'T RIGHT TO TALK ABOUT HOW YOU'RE GOING TO PLAY BEFORE THE GAME STARTS! COOL IT! Spend spring training drinking beer with the boys, put some extra pounds on. That can't hurt your game. Cut out all that dirty talk, and you'll be slamming home runs by game time.

If baseball was played as this backward manager wanted, we'd abandon major league games and stay right at home and watch the neighborhood Little League.

But how do you *know* if your batting average puts you in the Little League or if you are qualified to be a big league all-star? Locker room talk that you hear would lead you to believe that every man is a big leaguer. There is only one person that could critique your game, and that is your teammate.

Unfortunately, your teammate isn't likely to complain too much about your playing until it's too late. She'll probably play along until she is so fed up that she is willing to tear up your contract. She finally takes her grievance to the commissioner via the divorce court. Even then her complaints seldom come out honestly. When she says "mental cruelty" or "incompatibility," what she probably means is that you are a lousy lover—that you "struck out" in bed!

The trouble is that your dad probably flunked Little League, too, so how could he have been any help in teaching you the wonderful game of love? And your poor mother probably never knew there was a difference.

We never had the chance to peek inside anyone else's sex stadium until Kinsey rocked every bridge table and locker room almost twenty years ago with his in-depth research on our sex habits.

It was like breaking a log jam as far as bringing our "secret sport" out into the open. We've come a long way, baby, but we still have a long way to go!

The latest shock was the disclosure that in St. Louis a couple of great sex coaches named Masters and Johnson had opened a new training camp in which they taught troubled couples how to succeed in bed.

These couples had been making errors for years—and bad ones. Often he scored prematurely, sometimes not at all. Often he didn't get his partner warmed up for the game, or if he did, and they started to play, he popped out with a weak bunt, and the game was finished.

These couples had been so concerned about their errors that they gladly paid over three hundred dollars per day in fees and living expenses for a two-week training session. That's over four thousand dollars, if you care to add it up!

Batting averages and playing techniques are invariably improved by practice and expert coaching.

The fame of the Masters-Johnson clinic has spread so quickly that they may have to rename St. Louis "Sex Louis," and perhaps replace that enormous arch with a more phallic symbol of the town's virility!

A lot of women wish you men could cough up the money

157

to attend such a clinic and raise your batting average by learning to cut out all your illegal plays.

Phyllis Diller was quite blunt when we talked about the most common illegal play in the sport of sex—premature scoring on the part of the male.

"Hell, Debbie," she said, "it's not worth renting a motel room for less than thirty minutes—and I call anything less than fifteen minutes premature scoring."

So many of the women with whom I talked complain of this same problem.

One lady said, "I enticed him with a pitch low and away to keep him up at the plate. He reacted with "wham-bang home run . . . the game is over . . . and I'm still on the mound wanting to keep pitching."

"What is so maddening, Debbie, is that he thinks since he scored and I'm still wanting to play ball that I'm inadequate . . . hell, who wants to play a one-inning game? I don't think I'm asking too much if I want to go at least four and one-half innings; at least the game will count for something even though it isn't a full nine-inning game."

Think about this sporting event from a spectator's point of view. Say you are going to see Frazier and Clay fight for the world championship. You have paid a good price for your seat and you are ready for a good match. You are on the edge of your seat. Finally the bout starts and Frazier knocks Clay out *cold* in the first round and he is not able to get up again. You have sort of an empty feeling. You are really disappointed! You wanted more rounds. You just sit there for a minute stunned, wanting more action, and yet you know it's over. There will be no more clinches that night.

If you went back to see ten bouts and this happened every time, how many times do you think you would go back after that? You probably wouldn't go back at all!

Your partner in sex could probably go on and on about how she'd penalize you if she had the chance. She has a long mental list of penalty flags she's tossed down on your love-making techniques. (No, she hasn't told *you* about them, but she's told Ann Landers and Dear Abby; she's told her friends, her analyst, her gynecologist; she's told *Ladies' Home Journal, Cosmopolitan, Redbook;* and she's told me! If you don't believe me, read on and heed!)

Penalty Flag #1. *KICKING OFF BEFORE GAME LEGALLY STARTS:* "This may be forgivable the first time it happens. After the second game, if he kicks off too soon, it's approaching the unforgivable. After the third or fourth time it happens, it *is* unforgivable. By then I know it's *going* to happen and he turns me off, from his first kiss until he's finished with his sport.

"Oh, he apologizes, all right. He says that I'm just so great that it makes him so terribly passionate that he can't stand it, because I *overstimulate* him so. Well, can't he understand that I'm stimulated too?

"This is a point which he selfishly overlooks. I need satisfaction, too, but I just don't get it from this guy. All I get is a big letdown, frustration, and a psyche-out! I guess I'm just conditioned to anticipate losing the game before we even start playing."

Kicking off requires both teams to be ready if the game is played according to the rules. If your team kicks off before hers is ready, you must take the ball back and start again. The second time this happens, her team must be getting frustrated. Each successive time you kick off before her team does, the situation becomes worse, until your offending plays land you in the end zone and her team leaves the field, not willing to come back until your team plays with better sportsmanship. I'm willing to accept a few premature kickoffs in the game of sex, but if you aren't going to work at coordinating your timing with hers, I'm going to have to rule that another team should be allowed to play with her.

BACK TO THE CHALKBOARD TO LEARN TO AVOID KICKING OFF BEFORE THE GAME STARTS: My superstars report that men can avoid kicking off before the game starts in three ways.

"I had this trouble when I was in my early twenties," one

159

told me, "and on the advice of my doctor I started using two condoms instead of just one. This cut down on my sensitivity and allowed me to go a lot longer. Soon, the premature ejaculation just wasn't a problem any more."

Another of my thoughtful lovers told me that he practiced masturbation before beginning lovemaking. "It kept me from being unable to satisfy my wife," he said, "because I wasn't worried so much about my performance, I just concentrated on her. By the time she became excited, I was excited again, too." This same Hall of Fame member told me that whenever he concentrated *solely* on the object of his love—his wife—he was able to go much longer. "I think when men worry too much about their own performance or their own enjoyment, they're much more likely to come prematurely," he said. "I learned to almost disassociate my mind from my body, and to focus every part of me on satisfying her. Before long, I was making her much happier, and going much longer. Because I pleased her more, I found a new fulfillment in my own lovemaking."

Still a third way to avoid premature kickoffs is the "pinch method." Should you feel yourself near orgasm before she is ready, simply withdraw your penis. Then perform the pinch technique by giving the penis tip a hard squeeze for five seconds with your thumb and first two fingers, or have you wife do this for you. Then wait twenty seconds. Your erection will diminish slightly, but you will be able to continue your lovemaking for quite some time after reinsertion.

After you learn the pinch technique and continue to practice it, you will gradually increase the length of time you can continue lovemaking after insertion. One man told me that he had a longtime problem with premature ejaculation until he learned the pinch technique. "I used to be good only for about a minute or two after insertion," he said. "After four months of using the pinch technique, I had increased my time to over twenty minutes. I feel like a new man—and I guess I am!"

Penalty Flag #2. *FAILURE TO PROVIDE WARM-UP TIME:* "I'd never played in his ball park before. In fact, I'd used his bat only once. He insisted on going to his place. His lighting was hard to get used to, and I thought the turf had more bounce than mine. Before I met this guy I'd always felt that sex was beautiful and that anticipation was half the game.

"I told him I needed more time to warm up. He got antzy, led me right on to his playing field, immediately lined up for the kickoff before I could even get into proper field position, then kicked off, leaving me totally disgusted with his game plan. I'd like to protest his action to the commissioner—at the very least I can refuse to play with this guy again."

BACK TO THE CHALKBOARD TO PROVIDE ADEQUATE WARM-UP TIME: If she's unduly tense, you need to talk and cuddle for a long time. Let her get anything that's bothering her off her chest (to clear that lovely space for you!). To let her know you love her, take one whole night and promise yourself never to use the word "I" in your conversation. Never let her feel that the object of your whole evening together was to get her into bed! She'll be much more ready, much more in the mood, if she gets the feeling that sex with you grew out of her own desire, as well as yours. If you charm her and take plenty of time to wine her and dine her and to talk to her, she may arrive at the happy conclusion that the whole thing was her very own idea—and how much nicer the game will be for you!

Penalty Flag #3. *ILLEGAL SCORING:* "He was great in his warm-up, and I was mentally and physically prepared for a good game. Our game began in a marvelous way, with no tricks or surprise plays, as we moved up and down the field together. I had him figured as a guy who would be willing to let the clock run out, a guy who would never rush the end of a game.

"Then he purposely changed tactics and speeded up his plays without ever considering my playing tempo.

161

"Suddenly he kicked a field goal—and it was only the second down and goal-to-go, with plenty of time remaining! I had been ready to let him have the first touchdown, and he did that to me! Heck, what he did was worse than a premature kickoff. At least in a premature kickoff, I'm not all warmed up and hot to play the game out. But wow, with this guy, just as I reached the peak of my game, he goes for a field goal to satisfy himself, and I'm left lying there just staring at him.

"Then he rolls off the top of the pile and starts snoring. He's through—but why didn't he ever learn he could have scored six or seven points and made me very happy, too? Instead, he settles for his three points, and I never want to play with him again."

BACK TO THE CHALKBOARD TO MAKE UP FOR ILLEGAL SCORING: If you should score before she's ready, you've got to square the game with her. Try to avoid illegal scoring, of course, by doing what the coaches would do if they saw a player getting out of control. A coach would take such a player out of the game; if you find yourself getting out of control, simply withdraw for a while. Manipulate her with gentle stroking of her clitoris. Assuming she's quite well warmed up, give her oral sex, with your tongue beginning with gentle flicking motions over the clitoris, increasing to vaginal thrusts as she becomes more excited. Start thinking of pleasing her.

By all means, if you *should* score illegally, never roll over and go off to sleep. You owe it to your partner to manipulate her until she, too, is satisfied. This can be done manually, or by using one or two fingers to satisfy her, or by using a battery-powered "personal vibrator" available at most drugstores for less than ten dollars. The vibrator (made of plastic, in a shape that will remind you of your own anatomy) can be gently applied to her clitoris and moved as you would your own organ to bring her to climax. (One man I interviewed told me, however, that you'd be much better off not using the vibrator for a complete insertion. He swears that she'll discover that the

162

vibrator's a lot more satisfying than you'll ever be, since you just don't have the buzz of the batteries!)

By all means discuss your illegal scoring with her; tell her you want to make it ever so much better for her next time, and ask her advice. Working together, you two can become a beautiful team, but it certainly helps if both members of the team share the game plan.

Penalty Flag #4. *UNNECESSARY ROUGHNESS:* "From the very beginning this guy tries to show you what a superstar he is. To him, success implies unnecessary roughness, on every play, and when he's ready to ram across your goal line, you'd think he's using the 'old flying wedge' instead of a quarterback sneak. He'll roll on top, dig in, and never let up until it's over for him. Position to him is only one way—on top. He'll grunt and groan like a lineman to show you how masculine he is. It's like wrestling with a Brahma bull, and just about as passionate an experience."

BACK TO THE CHALKBOARD TO ELIMINATE UNNECESSARY ROUGHNESS: A woman who likes her love play a little harder than most will find plenty of ways of letting you know. She'll push her pelvis toward you, tighten her grasp around you, and, yes—even claw and bite when she gets very aroused. This can be your signal to increase the fervor of your lovemaking. Conversely, if you're with a woman who likes her lovemaking very, very gentle, you can give her a terrible turnoff if you start using caveman squeezes and thrusts from the very first. And by "turnoff," I mean you'll get her out of the mood so fast that she won't want a rematch for a long, long time. You've got to remember to start out with gentle caresses in all love matches and let your partner give you the clues as to how she wants the game to progress. This way, you'll never be penalized for unnecessary roughness.

Penalty Flag #5. *ILLEGAL CONVERSION AND SWITCHING ENDS OF THE FIELD:* "Ah! The innovative athlete! How great switching ends of the field can be if we both know that's the way the game is to be played. But this athlete never tells me what I should do or gives me a chance to tell him what I like.

"We were both enjoying normal play when suddenly he reversed the field—and expected me to help him score. I didn't really mind it . . . but what kind of game is it when I help him to the end zone and he doesn't ever reciprocate to help me?"

BACK TO THE CHALKBOARD TO STEER CLEAR OF ILLEGAL CONVERSION PENALTIES: This rule is easy; simply follow the golden rule and give unto your lover as you would have her give unto you. Never ask a woman to give you oral sex before you give it to her. If you want her to give oral sex to you for at least five minutes, give her at least five minutes of oral sex before you ask her! And remember that oral sex certainly isn't the *starting* point in lovemaking. It is used *after* you have aroused her, caressed her, and have gently run your hands all over her body.

Penalty Flag #6. *IMPERILING AMATEUR STANDING:* "I'm a big girl now about men and sex, but I still pride myself on maintaining my amateur standing. One thing is sure, I don't like being taken for granted. But during our series (and we played for about six months), he made me feel as though I played doubleheaders with different teams every night.

"I'll never forget one snowy night when we made love at his apartment. He should have gotten up and taken me home—I thought. But he didn't. Instead he handed me a couple of dollars for a cab and saw me to the door . . . on the eleventh floor of his apartment building. The very least he could have done was get dressed, take me down in the elevator, hail a cab, put me in it, and pay the cabby for fare and tip.

"Not this Sir Galahad! Instead, I leave his apartment, two dollars in hand, feeling not only like a prostitute but a horribly underpaid one at that!"

I guess things could be worse—some twerps don't even

think of paying a girl's way home. Even members of a losing football team get their transportation home paid!

BACK TO THE CHALKBOARD TO KEEP HER AMATEUR STANDING INTACT: Especially in our larger cities, it's often necessary that a man send a woman home in a cab alone. This is a bad enough situation, but to hand her the cab fare compounds the error. It makes a woman feel as cheap as a common streetwalker. If you must send your date home alone in a cab, give the driver her destination, ask him how much the fare will be, and pay the driver directly (with enough to cover an adequate tip). As further evidence of how much you care about her, tell her, "Be sure to call me when you get home, Honey," and ask the driver to either see her to her door or wait until she is safely inside. Never hand a woman money as though you were paying her off!

Penalty Flag #7. *BREAKING TRAINING RULES:* "Every time we've played, he's guzzled about five drinks to get himself warmed up—then it's fumble, fumble, fumble.

"Just inserting his bat is a problem, because he can't tell my first base from home plate, all his swings fall short of connecting; it's pathetic. By the time we play, he's so numb that he can't even tell how *he* feels, much less how I *might* possibly feel. It's bad even in our warm-up—I mean, have you ever had a guy try to French kiss your left nostril?

"The only sensation this guy ever gets is a terrible hangover."

BACK TO THE CHALKBOARD TO PLAN A BETTER TRAINING PROGRAM: If you drive, don't drink. If you must drink, don't drive anything (especially your playing equipment). Instead, sleep off your stupor and try lovemaking afresh in the morning.

If you tell yourself you're drinking to cover up your bad plays, you're only exchanging one bad play for two. Face up to your problems (premature scoring or whatever) and follow the suggestions above. Some men say one or two drinks help them go longer. Well, the drinks *do* slow you down. But remember that if you consume four or five drinks, you're not only slowed down but close to down and out as far as your successful game is concerned. Can you imagine any major league athlete drinking before an important game?

Penalty Flag #8. *CAUGHT PLAYING WITHOUT FIRST LEARNING TECHNIQUE:* "I think it's only fair to assume that a girl will get a player who's at least evenly matched as far as playing ability.

"I mean, I have spent a long time learning how to play my game. I've learned to use warm oil in my mouth to make his oral sex more enjoyable. I've learned to put my legs on his shoulders if he wants to insert farther. I've learned to touch his testicles during intercourse—I've read every book on the subject of sex, so that I can make the match as great as possible.

"Then along comes this Little Leaguer who thinks he's the world's greatest lover, and I find that his hottest trick is about as exciting as playing Spin the Bottle. Well, I quit playing that game when I was nine! I expect quite a bit of sophistication from the man I play with now."

BACK TO THE CHALKBOARD TO MASTER BASIC PLAYS: Learn to cuddle and to huddle properly. Learn to switch ends, to play new positions, to give her some surprise plays.

Learn to stroke the underside of her hips. Learn how sensitive the front of her pelvis is. Learn to apply soft pressure with your entire hand as you massage her entire vaginal area, slowly but with increasing pressure. Learn to use a vibrator to gently tease the area around the vagina, tugging gently at the labia (lips of skin surrounding the vagina) as you do so. Learn to take your thumb and forefinger and give the labia teasing little tugs. Learn to be patient and wait until her vaginal secretions appear . . . a sure sign that you have made her body react to yours and create that special chemistry which is the essence of lovemaking. Best of all, learn to make your mastery of her feelings your basic technique. Devote one hour just toward pleasing her, toward evoking new reactions from her. The time you spend will be repaid to you tenfold in the new pleasure you'll derive from the act of love.

Penalty Flag #9. *PLAYING WITH IMPROPER EQUIPMENT:* "I assumed before we started to play (and he instigated the game, for heaven's sake) that he'd be sure the balls were properly inflated and that the bat could get the job done.

"I know *my* equipment was in order, and ready to give a good and satisfying game. But he seemed to think his responsibility ended when he delivered the deflated balls and a weak bat to the playing field in his overweight van. What a disappointment; I called off the game when I realized he couldn't make it through even the first inning!"

BACK TO THE CHALKBOARD TO IMPROVE YOUR EQUIPMENT: If you are not physically up to par, cancel your game. If impotence is your problem, please get a thorough physical examination. If no physical causes for your problem turn up, see a psychologist or a psychiatrist. Getting new teams to play with you every night will certainly not solve your problem—but will confirm your failures. You owe it to yourself to take every step available to make your recovery possible.

Penalty Flag #10. *CRUDE PLAYING TACTICS:* "Mutual friends had given him my name, and he invited me to have lunch with him and asked that I come up to his room for a drink first. I went—and he tried to get me right in bed with him. Ugh! I don't know how he got the idea he could even try such a thing. I was mad, and hurt, and in a miserable position. The worst part of it was I had the feeling he didn't even like me very much. We finally did have lunch, after he'd tried every rude tactic in the book to get me in bed with him. He was sullen and surly. He never called me again, which didn't bother me in the least. But I certainly called the couple who got us together and told them what had happened."

BACK TO THE CHALKBOARD TO LEARN BETTER MANNERS: Any man who tries to leap into bed with a woman he barely knows is a pretty awful player. Any man who tries to lure a woman he doesn't even like much into bed is worse—an alley cat for whom sex has no real meaning at all. A player who would do such a thing is saying, in essence, "I don't like myself, so it isn't necessary that I find a partner I respect and care about." A player who uses these tactics is waving a flag that spells out trouble for him. He's in need of professional help to allow him to understand why he's behaving as he is, to improve his feelings about himself, and to restructure his self-worth.

When you have eliminated the chance of even one penalty flag being dropped on your game of love, you'll be able to rank with the superstars you'll read about in the next chapter.

And when that happy time comes, you can just erase everything on your chalkboard.

Don't ever throw the chalkboard away, though—save it for scheduling all the love matches she'll be begging you for in the wonderful days and years to come.

Chapter 7
The Superstars Of The Sex League

In my thousands of feet of taped interviews, I ran across several dozen men whom I would rate as superstars in the great game of sex. These men have spent years and years learning techniques of playing the game of sex, and I have put them in my Hall of Fame. (I say "my" only because they were nice enough to share their experiences via my little black Norelco. I don't doubt that the women who are in these fellows' little black books would give rousing cheers when asked about their playing abilities!)

These Hall of Fame members are just like you . . . in terms of basic equipment.

They are equipped for the game with the same set of reflexes, muscles, and tissues and very similar bat and balls.

At sixteen, you and my Hall of Famers were probably pretty much alike in your playing abilities.

I've learned, after some 150 interviews with highly touted lovers, that not one of them consciously started on a career of lovemaking . . . but now they rate the all-American list.

If you are wondering why you can't satisfy your wife or

lover, listen to this interview with Britt, one of my top ten performers.

The questions are mine, and the answers his.

Britt is rugged, good looking, age forty-four and a successful businessman and outstanding sportsman.

QUESTION: You have known so many women—what about the frigid woman?

ANSWER: The what?

QUESTION: The woman who cannot respond to love-making—the one who fails to reach orgasm.

ANSWER: I've never known any such woman.

QUESTION: But most women complain that a man doesn't know how to turn them on. How did you learn to make a woman happy?

ANSWER: My situation was, it was luck. I was eighteen and doing diving exhibitions at a hotel in Miami when a bellboy delivered a note to me saying that a woman wanted to take me to dinner.

I asked around and found out that she was sixty-two— but I went out to dinner, and then to her house. Sure, I had feelings of—fear, embarrassment.

As it turned out, I stayed with her for two years. She taught me everything I could have learned about sex, and this was twenty-four, twenty-five years ago.

QUESTION: What did she teach you?

ANSWER: Well, I'd go to the house twice a week and the servants would get out of the way. We did the soapy shower routine together with scented candlelight in the bathroom, and afterward we would lie down together on her satin sheets and do the "Mazola Roll". . . .

QUESTION: The Mazola Roll?

ANSWER: You know . . . just lubricate our bodies with cooking oil and feel each other, and get with it on her satin sheets, and hear the soft sucking and slipping of our bodies together . . . and our hands gliding over each other's bodies. It was sensuous.

(Note: And anything polyunsaturated is bound to be good for you!)

170

Britt graduated, at twenty, from his "stadium" in the home of his widowed sixty-two-year-old mistress.

He still speaks of her with an uncommon mixture of tenderness and admiration, much as a child would speak of a beloved teacher.

"I have a son," Britt said, "and I only hope that he finds an experienced older woman just as I did. That way, he'd become a pretty nearly perfect lover. He'd learn all the things that would make his woman want him, and only him, for the duration of their lives together."

Let's add a note on Britt. Divorced, but with great concern for the welfare and future marriages of his children, at forty-four, Britt still seeks a future wife-lover. He still seeks a woman for marriage . . . one who would bear him children but would make the children the by-product of their union, rather than the *cause* of it.

In the meantime, his bachelor pad is, yes, a sex-stadium. Britt has learned all the things that turn a woman on.

In Britt's case, the stadium is a high-rise bachelor apartment. The rookie player could learn a lot from a tour of it. It's a chic, elegant apartment, with a sophisticated stereo system, an immaculately clean kitchen (with real *food* in the refrigerator, and a couple of bottles of wine kept chilled). There are lots of books (obviously read) scattered on the long coffee table. There are some interesting oil paintings of nudes, all abstracts. There's a fireplace with a bearskin rug in front of it, and a large balcony for looking out over the lights of the city.

The bedroom? The king-size bed is topped with a gold canopy to match the velvet spread . . . and the canopy frames a smoky-glassed ceiling mirror in the exact dimensions of the bed. The bedside table holds an electric vibrator (purchased at a nearby drugstore for $9.95) with four attachments to enhance the art of lovemaking. Britt always has a supply of birth control devices on hand. "Sure, most women are on the pill. If they're not, they'll usually tell you—you don't have to ask. But if for some reason I wasn't sure, I'd certainly find out before we started making love."

KNOWLEDGEABLE PLAYERS
KNOW THE SCORE

Scoring is a pretty wonderful feeling—but it's a shallow victory, my Hall of Famers agree, unless you help your partner to score right along with you (heaven), or before you (thoughtful), or certainly after you (a must, if you expect to have rematches in the future!).

My Hall of Famers put a lot of thought into their lovemaking. They don't keep their high rating with women by being selfish, you know! Listen to Bob W., a California stockbroker, happily married for eleven years. I consider Bob's philosophy a classic that all lovers should reread a couple of times each season.

"An analogy I would make is that if I were building a fire, I could do it several ways. I could just crumple up paper and toss on a match, and the paper could blaze up all at once, and then it would be over. This is the way some men are about sex. They just crumple up a paper, so to speak, and the blaze is over immediately.

"On the other hand, if the man builds a careful fire, he has to use little twigs to build it, and he has to put the logs on carefully and build a great rapport to keep that fire going. Then he'll have a fire that will keep blazing for a much longer time, and it will be a much warmer fire, and *this is what a woman wants.*

"The quick crumpling of paper and the fast blaze isn't confined to new relationships, either. Men who are married, even though they can be very much in love with their wives, may still light the crumpled paper for the first blaze—and I know they usually feel pretty guilty about it, too."

While we're talking about attitudes, let's discuss the fine art of getting her ready for the game of sex.

172

My all-stars agree that sex is overwhelmingly in the head, that creating a mental attitude, a romantic setting, a sense of *her* wanting *you*, makes your mutual success at the game much more likely.

Said Marty, a sought-after man-about-town:

"Intercourse—and my scoring—has to come about as a natural sequence of the events of the evening. It can't be the cause, or the girl would feel I was using her. I always try to make sex just happen, just come about naturally. If we start on the couch, or on the lanai, fine—we can stay there, for all I care. But it's a smooth, natural way of doing things. Women appreciate this."

Terri, a former Playgirl-of-the-Year now employed at the Playboy Club, seconded the idea that good lovers never let the woman feel sex is the foremost reason for their relationship.

"I'm living with a man now," she said. "Every Monday he sends me a dozen sweetheart roses delivered to the club. It makes me feel loved every minute of the week and makes me feel proud. I think it would be much better if men made gestures—little thoughtful gestures—instead of barging home or to a girl's apartment just to have sex. Give for the sake of giving, just to show a woman that you care about her. I think it would make a woman feel more receptive all the time."

Most women hit the peak of their sexual desire several days to a week prior to their menstrual period. Unfortunately, this is the same period that often finds them rather moody, highly sensitive, and irritable. The trick, of course, is to let her know that you want to play ball with her every day of the month, yet keep her aware that you think of her not just as a sex object but as a highly desirable woman.

If you're dating a girl who you feel is "hard to get," turn tables on her. Continue to be your well-dressed, physically fit, considerate self, but try the same approach that Jack L., an easterner who inherited millions, uses with his women:

173

"I don't play games. But I date this way. First, I take a girl out because of physical attraction. Then, over dinner, we talk, and I find out if there is anything in that pretty head of hers.

"I can't stand girls who *play* hard to get, but I know there are some who *are* hard to get. Often, it's her own hang-ups that create a situation. With this type, I take her home and maybe just give her a warm caress on the shoulder, or hold her hands in a warm clasp and tell her it was a very pleasant evening.

"If I really like her I make another date, and the same nonaggression is repeated. Maybe I send her flowers.

"By the fourth date, she either figures I really care about her (which I do, or I wouldn't have wasted so much time) or that I'm a queer. Either way she trusts me. Only then are we ready to begin lovemaking."

D. K., a highly respected obstetrician-gynecologist in the Denver area, has long been in a position to counsel women—married and unmarried. His reputation is such that women make appointments just to talk over their psychological problems.

Listen to this.

"Of course sex is a basic drive. But if a man wants a really good relationship with his wife, he's got to realize that her days at home with one or more children are one continuous demand on every minute of her time. A woman with a small baby—her life looks so easy to the man. But by the time the baby gets down for a nap, the woman has to do all the routine things—cooking, washing, housecleaning.

"So many of my patients come to me when their children are little and voice the fear that they have become frigid.

"A husband can win her back by helping just a little—offering to load the dishwasher while she flops on the bed and reads a magazine. He can let her know he appreciates her efforts at home every day."

Most women are stimulated by pelvic massage in sex.

174

You learn to use your pelvic area just as you want her to use hers. Men often complain of women just lying there doing nothing. When women do respond, they do so quite often by thrusting the pelvis forward. This is not to be confused with just lifting the hips. There is extra emphasis on yielding the pelvis. If a woman is in the superior position, you can learn to lead up to her with the pelvis. The pelvis leads and the rest follows. Holding the pelvis against her firmly is also quite stimulating. This is that extra little giving that gives her just a lot more pleasure. Learn to be proficient in this technique because you'll want to use it in many other positions.

It is recommended that you work on one position until your control is at a maximum and you can give her maximum pleasure in that particular position. Learning all the positions in the world and doing them without the feeling of love and desire to give completely is worthless. You should want to be pleased in sex by having the opportunity to please your partner. And you should carry the idea of "pleasing" into every phase of your relationship.

Says Keith R., home developer:

"I feel that I have to understand how much her hangnail hurts. That's caring. Or how tired she is after she has worked all day at her job. If you feel you can really sense the same emotions that the woman does—only then can you give her the sort of love in bed that makes you both feel like a million bucks."

IS THE WORLD YOUR STADIUM?

I don't care if the invitations have just gone out for your silver or your golden wedding anniversary, if you're just starting your honeymoon, or if you have decided to have a fling with that cute new secretary.

If you want your love life to zing, you've got to use your imagination! Use every bit of that wit of yours to make your love life fascinating, tonight and forever!

Hall of Fame men are masters at this. They keep their lovemaking alive!

Betty P., a twenty-five-year-old nurse, told me that her husband scrutinizes every motel room as they check in— but not for color television, stacks of clean towels, or any of that trivia. "Oh, Bruce would probably complain if the bed were downright lumpy, but the motel *chairs* are what concern him most. We were never really good lovers until we started using a chair," says Betty. "I learned that he could get better penetration and that he could position himself in a way that would allow me to move in many directions until I found the most stimulating one. I like a chair that has a crossbar so I can position my feet and work my pelvis with freedom.

"It got to be such a thing with us that after we were married, when we were buying things for our first apartment, we'd both get terribly passionate in furniture stores, just at the sight of a chair with the right structure for sex!"

Al, one of my Hall of Famers, kept emphasizing how important a headboard was in lovemaking. His lover agreed. "I like a man to be big and husky," said Sue, "but I don't go for that nonsense that you don't feel their weight when you're making love. Of course you do, and if the man is one hundred pounds heavier than the woman, its going to crush her. Al uses the headboard to brace his feet, and it takes a lot of the weight off me when he's on top. Plus he uses the resistance to help give him better control with penetration. I like to have him tease me a bit, not going all the way in. After he has done this for a few minutes to then go in all the way . . . and just hold it deep inside me. Bracing his feet certainly helps this action. My bed doesn't have a headboard, so if we're at my apartment, we either use the end of the couch or get down on the carpet so that Al can brace his feet against the wall."

If you haven't discovered the sheer joy of communal bathing or showering, where have you been? Swimming pools are fun, too, but if your playing field doesn't have one, a smaller body of water will still make do just as nicely.

Granted, this might not be the answer for your first-time match with your conquest, but please don't miss out on this most wonderful prelude to lovemaking.

Just listen to what a few of my Hall of Fame members had to say about fun in the water:

"The first time my wife and I tried bathing together was just sort of an accident. She was in the tub, and we were getting ready to go out. I went into the bathroom to get my razor and—well, there were a lot of bubbles, and suddenly I was just in there with her. We took turns sort of massaging each other with soap, and one thing led to another. We were an hour and a half late for the party—but who cares?"

He went on to make other good points:

"I definitely think what the psychologists say is true, that if you're going to give love, you really have to love yourself first. Well, I don't have a fetish about being clean or anything, but who in the world doesn't love himself more when he's fresh from the shower and feeling really clean all over? I mean, even if you've done nothing strenuous since your morning bath, it's relaxing and refreshing to shower and a lot more fun to do it with your woman, too

"There's something about warm water pouring over bodies that takes the sting out of any arguments or tension. You suddenly find yourself relaxed with one another, and all your pores are open."

Hints from the pros also include the following (you don't have to have Handel's *Water Music* on the stereo, but at least give the following a whirl): The pros say that a lot of women love the shower-tub idea, but for heaven's sake don't try it under the same bright fluorescent lights *you* use for shaving. If you've never thought of your bathroom as the sort of place to be suffused in candlelight, get with it. A scented candle isn't a bad idea, either. And be sure to have plenty of lush, thick towels ready for afterward.

Come on out now, even though the water was fine!

Before we move on to the next inning, let's take a brief

tour of some of my all-Americans' favorite getaways for the game of sex.

Two delightful suburbanites whom I interviewed said that they had felt trapped one Thanksgiving weekend with an overflow of houseguests and had been unable to feel that lovely "all-alone" sensation for some time. They hit upon their enormous walk-in closet and made it their "room with no windows" for several nights, substituting a scented candle for mothballs. The experience, they claim, literally brought their stale lovemaking *out of storage*.

A highly successful southwest builder (so successful that he's retired at the ripe old age of thirty-four) reported that he and his wife find camping out in the woods a highly stimulating way to make love. The two children are left at home with their sitter, and the couple drive out to a secluded spot, toss out a blanket, and get really turned on. (Oh, yes, their camper is a beautiful twenty-six-foot land yacht just in case the weather turns bad.)

A well-known athlete said he and his wife love to go swimming in a lake or stream, way off by themselves. "Lovemaking in water is one of our greatest thrills," he says. Great on a moonlit night.

A TV celebrity and her husband told me that they like winter skiing, but from what I gathered, the time they spend on the slopes is minimal. Rather, they use the enclosed (but of course) gondola lift, hang a Do Not Disturb sign on the door, and spend the fifteen-minute ascent making love.

It all began when she and husband were burning up the ski slopes one sunlit weekday morning and they suddenly realized that their mutual warm feeling came from something far greater than their fur-lined parkas. They pulled up in the shelter of a thick pine grove and assessed the situation.

HE: Shall we?
SHE: Here?
HE: Why not?
SHE: How?

HE: Mm . . . I think you better start by taking off your skis!

My friend, still dubious about the possibility of passion in the snow, immediately agreed that both their chances were better if they removed their skis. After considerable maneuvering, they arranged themselves in a time-honored position for lovemaking. The fun had just begun when they were interrupted by a stray skier! The climax of this is that there was, alas, no climax. Red faced, they strapped their skis back on and made their way back to the lift, when they were again overcome with that overpowering urge to make love. The ascent was a long one, and this time my friends had ample time to warm up and enjoy the full heat of their love. They discovered, incidentally, that the ski gondola seats made possible a lovely variety of interesting new positions. When the lift reached the top, our skiers were feeling that lovely warm afterglow that's a bonus of good lovemaking. They stayed in the lift, cuddled all during the trip down, and returned to their nearby condominium for lunch and a nap—after which they felt rested up enough for another session on the slopes.

After this "happening," they told me, they used the ski gondola quite often, and claim that anyone who reserves it for only skiing is missing a fun sex spot.

Still another lovemaking idea comes from Tom R., a Louisiana oilman. Tom and his young wife like to pack a picnic lunch and take off on their horses for a ride through the lush wooded area north of their Baton Rouge home. They sort of let things happen and give their horses their head until they find a pleasant secluded place for their lunch. Tom swears it's the pelvic stimulation that his wife receives from the ride and says all he has to do is cuddle with her a bit and she practically rapes him—in a very wifely, proper way, of course! I wonder if the same holds true for motorcycling—and if that's why all those couples I see riding double seem to be having so much fun.

Chapter 8
What's Happening In Your Bedroom?

If ever you should find yourself at a total loss as to what really turns a woman on (and every single woman is different, you know), you might try a trick which my favorite psychology teacher taught me.

"Pleasurable visual stimulants," said Dr. Paul Knott, a psychologist, "cause the pupils of the eyes to dilate. This is an involuntary reflex, one that is impossible to control." Would you believe that Italian women, as early as 1450, dropped atropine in their eyes before their lovers came, in order to heighten the appearance of passion? In fact, atropine was then known as belladonna (literally, "beautiful woman")!

As further proof of this I remember a psychological test we performed which involved two "identical" photographs of a beautiful young woman. One of the "identical" photos had been retouched so that the pupils were dilated. Male students were asked to select the more attractive of the two photos. In almost every case, the picture with the eyes dilated was selected—but none of the boys could say why.

The reasons for their choice could revolutionize your love life!

If ever you feel you're about to flunk out with your favorite female, try "playing doctor" to find out what really turns her on.

All you have to do is rent one of these so-called underground movies and arrange a cozy screening for two.

Then, armed with your penlite and optical reflector, simply observe the pupils of her eyes as she watches the movie! Pupils dilated—and you've got your signal to follow up with the same action she's viewing on the screen.

Pupils contracted—and you know that whatever action she's watching is a complete sexual turnoff.

Of course, if you can maneuver such a rig for the purpose of testing your lady's love reactions, you're a pretty shrewd operator in the first place.

But if she doesn't mind "playing doctor" with you, the game might be fun. After you learn her reaction, you can let her be the doctor and watch *your* reflexes. Don't mistake the effects of a darkened room for increased arousal. We may try to hide a lot of things about our psyches, but our pupils simply cannot tell a lie!

STRATEGY FOR THE HUDDLE
AND CUDDLE

Human females are the only female beings on earth who can experience climax, and they should. This God-given gift should have every possible opportunity for enjoyment.

"Warm-ups," said our Playboy bunny, "are everything.

"If more men would learn how important this part of lovemaking is, the rest would pretty much take care of itself, I think.

"Some guys think they're more virile if they give a woman a very hard kiss and crush her rib cage—then it's off to bed.

"Other types I hate, try to kiss every finger, every toe. You know what really turns me off? I've had men lick my fingers, thinking they're so sexy. To me that's just a bunch of slobber!"

(However, to each her own. The same suburban house-wife who found making love in her walk-in closet great fun told me that she becomes very aroused when her husband gently licks her fingers and toes. "He *doesn't* slobber, though," she laughed, in answer to my question.)

All the women I interviewed agreed that a man's attitude toward her body is very important.

"I guess every women reacts to how a man feels about her breasts," said Sheila, the dark-haired stewardess. "I get very aroused when a man first caresses my breasts very gently with his hands. He can feel my excitement, and that gives him the cue to move down with his mouth—ever so gently. Some women, I guess, like it harder, but I think that type would tell you; she'd let the man know. For instance, the woman who would initiate the hard French kiss is probably very passionate, and might want to be handled more roughly. But no man could go wrong by just stroking and petting, by getting the woman aroused very slowly, and the breasts are the most sensitive area for most women before intercourse starts."

A petite blonde nurse whom I interviewed likes to have her hair stroked, then her neck, then the breasts.

"I love the feel of an electric vibrator around my breasts," she said. "But a guy has to know what he's doing. It has to be gentle, very gentle. He can even go down and use it on my clitoris, but he has to have learned how to do it right. Very gently, very slowly, but—it's great.

"Some men think it's sexy to blow in my ears," she continued. "That turns me off completely.

"But my lips are full of feeling. A soft, massaging kiss, with him brushing his lips back and forth, back and forth, across mine, can almost make me come. It's the same with my breasts, my navel. I like a man to use his tongue, but with soft, sort of flicking, motions."

The KEYWORD with the women I talked to was "gentleness."

I would like to add one thing, a factor in lovemaking that is so very important. Don't *you* have an area of your body that you feel doesn't quite measure up? Most women do, too.

With women, it's usually an area that they feel needs some improvement. It might be a slightly underdeveloped bust . . . or a waistline that isn't quite slim enough . . . or stretch marks from the last baby.

While you're caressing a woman, take care to sense out what *she* is feeling. If she responds with a coo and gently maneuvers her body closer toward your loving hands, then you're "tendering" an area of her body that makes her feel like a beautiful woman. But if she tenses and moves away from your hands, you have your clue—she may feel that she's insufficient or overly ample in a certain area, and would rather you move on. Quite often she will move into a position that won't allow you to touch an area she doesn't want touched. Get the message and stay away!

After you get to know her very well, if you realize she has an area she's very sensitive about, you can turn her sensitivity into a plus factor in your loving. "I know Kathy feels an abdominal scar she has is unsightly," said one wonderful man I interviewed. "I always try to give the area a gentle, gentle kiss. It's like saying, 'I love every part of your body, everything that's part of you.'" Isn't that nice? His kiss tells of his love and concern, more than words ever could!

If your feelings of love are real as you begin the fore-play of sex, she will signal you at each base along the way and will be more than ready to go along with you.

One of her first signals will be a pulling closer to you, a maneuvering of her pelvis so that it is thrust forward toward yours. Most likely, you will actually feel her body heat increase and her breathing become slightly more rapid. She may manipulate her body so that she is almost astraddle one of your legs, and will push her pelvis against

184

the leg. But the most important signal of all is that when she is fully warmed up for the act of love, the area around and within her vagina will become quite moist as her body begins the secretions that lubricate the vagina to allow ease of insertion and sufficient moisture during intercourse.

If you watch for these signals and wait until she gives them to you before rushing into the game, she will be more ready to go along with you as you begin insertion and the careful, delightful, and exciting trip *together* toward the climax of your game.

BEGINNING THE INNING

As for when you should cease your warm-up and begin insertion, I think the cue should come from your teammate!

The primary rule for you to remember is that if you let her call the plays, in response to your slow and gentle stimulation of her body, the results of your match will be major league stuff!

So follow her cue and prolong the wonderful excitement of foreplay until she tells you she's ready.

What positions do you both play? Here, I'll go along with the "doing what comes naturally" group! Once again, she'll tell you if she's pleased or displeased with a certain position, so take your cues from her.

It is that beautiful moment now . . . her field is ready, and your bat and balls are ready to play.

Insertion should be easy, but firm, a thrust that is at once tentative and affirmative. You are, in the first contact, giving her the option of calling the tempo of the game.

Slow and easy does it . . . take your cues from your partner, and your game will be a winner for *both* sides.

After insertion, your home run *could* come at any time —but if you make it too soon, your poor partner will still be getting to first base—just as you're ready to retire from the game! And this is *no* way to play ball.

My Hall of Famers say they are able to sustain an erection for a *minimum* of thirty minutes, and up to three hours, or as long as is necessary to satisfy their partner. One says, "Achieving orgasm simultaneously is the point of the whole game, as far as I'm concerned."

Another Hall of Famer told me: "I used to think that achieving orgasm simultaneously was the point of the whole game; in fact, it seemed to be so important that I worried when I couldn't get us both coming together. Then, as I got more experienced, I decided this wasn't so . . . she can have hers, and I can have mine, and the timing doesn't matter that much . . . you can double your pleasure when the man and the woman come at different times."

Another great lover (a Philadelphia businessman and father of a large brood) admitted that he had a "bad time sometimes about three or four minutes after insertion. Sometimes I feel as though I'm going to come at that point, even though I can damn well tell my wife isn't nearly ready.

"It takes some concentration on my part to get over that period—or, better yet, nonconcentration. I can think about the stock market, about sheep jumping over fences —but I *have* to divert my mind from sex completely. This usually helps me get over the bad period—then, after that danger point is past, I can go for hours until my wife's ready."

Most of my Hall of Famers had tried the "pinching technique" of prolonging intercourse when necessary to satisfy their partners.

"It works—I know that," reported Bob S., a Denver lawyer. "We were having some sex problems, and our doctor taught my wife to place her thumb and two fingers on my penis and give it a firm pinch. A doctor ought to teach this to every woman, I think. When I'm in danger of going too soon, I tell my wife and withdraw, and she gives the tip of my organ a hard squeeze. I can go for as long as we both want after that.

"After we tried the pinch trick a few times, I got over my fear of failure. We hardly ever use it now because I can go as long as I have to."

For Britt, who learned all his techniques from his sixty-two-year-old mistress, this forestalling of ejaculation method is one that he does himself. "The woman is hardly aware that I'm doing it, and I think that's better," he says.

We asked about anesthetic creams, which have been widely used in marriage counseling clinics to forestall premature ejaculation. Britt said he had found them relatively ineffective—compared to the pinching technique which he has used for twenty-six years.

"I think that sex is about 99 percent in the head," emphasized another Hall of Famer. "Men with hang-ups usually are the ones that don't take the time to get a woman really warmed up or don't really care about the woman's satisfaction. The act becomes much easier when you forget yourself and simply play the game of seeing how completely happy you can make the woman.

"If you go for twenty minutes, or two hours, what's the difference? The main point is that she'll tell you in no uncertain terms when she's ready, and you'll make the climax together. God, that's a wonderful feeling."

But don't forget the job isn't finished even with a great climax. Afterward, a guy has to take the time to let his wife gently come out of it. They need loving, caressing, petting sex, too.

"One thing that really turns me off is to have a woman rush off to the bathroom after lovemaking . . . as though she was getting rid of something unclean. Women feel the same way, too, I think—at least that's what a girl once told me. The 'drifting' period in bed together after a good climax is beautiful, an intimate needing. It's a *time* to feel each other's bodies relax, not a time to be leaping off to the john. What more wonderful feeling than to fall off to sleep deep inside a woman?"

ON REVERSING THE
PLAYING FIELD

Most women are content to play the game by the standard rules, with the classic techniques. Men are often more likely to be adventurous in their innovations. However, almost everyone has a pretty strong view on oral sex. Unfortunately, it is usually a silent viewpoint!

Sheila, I think, spoke for most of the women I interviewed on the subject.

"Oral sex is fine; I enjoy it, but it all depends on how it's done and with whom. I have to feel that he is giving me something, too, before it has any meaning for me. In other words, I want to be satisfied orally, too.

"The worst thing is when a man pushes oral sex. Some women simply don't like it. Others probably could learn to like it with the right partner.

"But you get one of these crude guys—it's unbelievable. They'll take a girl's head and *push* it toward the penis with their hand. That's revolting, and thoughtless. It's a complete turnoff."

In one of my taped interviews, Garth L., thirty-eight, a West Coast studio executive, talked about oral sex.

"I think there are two kinds of oral sex. First, there's the kind that's just sort of negative. It doesn't disgust, but then it doesn't really do much for the people involved, either. In oral sex like this, I feel that the woman—or the man—can take it or leave it. They are probably doing it for their partner's satisfaction, but they get almost nothing from it themselves.

"Then there's oral sex between two people who really, really care about each other, and who have had a longer relationship. This kind of oral sex is tremendously intimate, and it's a marvelous addition to lovemaking.

"In this mutual situation it's a great way of having intercourse because it's the ultimate in intimate sharing."

188

To give her the kind of oral sex that will make her understand how beautiful it can be, you will use your tongue in swift flicking motions in her clitoral area. You will gently suck at the labia that surround her vagina, and gradually increase the pressure with your mouth—just the way you sucked when you were an infant at your mother's breast. Continue this, and you will feel her getting more and more excited until you bring her to orgasm.

A psychiatrist who specializes in sex problems had this to say:

"As far as oral sex goes, it's a thing about which many people have hang-ups. Women tell me they like to be stimulated orally *themselves,* that they find it highly pleasurable. But they don't like to give oral sex unless they are stimulated orally by their husband.

"Generally, on the subject of oral sex, I can say that no one has ever *enjoyed* it unless they've *tried* it, just as with heterosexual sex. If both the man and the woman are satisfied with oral sex, and perfectly happy in performing it, I wouldn't worry about their relationship at all. Remember, 'abnormal' is only what a certain culture *says* is 'abnormal.'"

Most of my interviews bore out Garth's viewpoints on oral sex. Many of the women I talked with claimed that participating in oral sex *became* highly pleasurable to them, although most said they had to be highly motivated and very much in love with their partner before the act became meaningful.

I quote stewardess Sheila again:

"My chief feeling about oral sex is the same as it is about lovemaking in general.

"I think it is very important that both men and women make only those demands on one another that they are willing to give. Specifically, when we're talking about oral love, I have found from experience that *every* man thoroughly enjoys this type of lovemaking, but that very few of them are willing to reciprocate to the woman. I don't think they are conscious of this, or maybe they feel that they are making up for this lack in other ways. But I feel that if I give oral sex to the man, he certainly should give oral sex to me.

"I've learned how to make it good for a man. I never let my teeth get in the way, and I use my lips to stimulate him just at insertion. I hold some warm liquid in my mouth—men like that. It can be alcohol, or oil, or just plain warm water, but men are stimulated by this.

"You know what the big turnoff is for most women when an inexperienced man wants oral sex? It's using real force to make her do something she doesn't want to do.

"If it isn't pleasant for her, she'll never do it again. And if a man always wants it, and she doesn't, he'll turn her against it forever.

"She'll get so conditioned to having to do something she doesn't like that she'll be turned off on *all* his types of love.

"But if *he* gives *her* oral sex *first*, then she'll understand how wonderful it is, and she'll be willing to give it to him because she'll know how it feels."

AFTER THE GAME IS OVER

The players who are brought together in the stadium of love have very different feelings about the game they play.

The male player is often ready to take his deflated bat and balls and retire to his side of the bed for a good long snooze.

The female player, however, usually wants to do the reverse of the warm-up. She wants to unwind!

She'd prefer to talk some more about their game, to have a kind of hushed verbal replay, and to cuddle as she drifts off to sleep still thinking about how pleasurable the game was!

One of the most beautiful parts of the game of sex, for the woman, is to have this cozy postlude to the sport. It means a lot to her. (Can you imagine how the quarterback who carried the ball over the line for a winning touchdown to win the game would feel if he was sent to the locker room for a solitary shower and then banished to a silent bed in his room?)

What do people want when they feel on top of the world?

They want to share their feelings, that's what!

And, after all your training in how to excite a woman before and during the act of sex, if you fail her afterward and don't understand her need to tell you how wonderful *you* are—you are missing her very private press conference full of praise for you . . . after the beautiful game that she and you played and won together!

Chapter 9
The Sex Machine

When you run out of plays, when the chalkboard offers no new ideas, when your love game has a bad case of the blahs, modern technology provides a nifty glowworm which enables you to experience a perfectly electrifying glimmer of love, "to turn on the AC and DC," as the song says.

Science has come up with a device that will give you sexual stimulation—via electrodes implanted in the hypothalamus area of the brain (the area that houses our sex sensors). The doctor who gave me information on this told of seeing a man hospitalized for the implantation. From what the doctor said to me, the patient had to be the happiest hospital patient in medical history:

"He was lying in bed, with this big grin all over his face, and he was so pleased with the electrode stimulation that he couldn't have cared less if he ever had a visitor! On his bedside table along with the usual hospital paraphernalia was a little black box that he could switch on as he wanted more stimulation.

Here's how the doctor explained the working of the electrode stimulators:

"The electrode is a tiny wire inserted through the skull

and brain tissue to the appropriate area of the brain," he told me. "The wire has to be in just the right place." (Like some other sexual stimulators I know.) "After the subject's hair grows back, there's no evidence at all that the electrode has been implanted.

"Exactly the right level of radio-wave stimulation must be used. The level is very low, otherwise the stimulation would be painful.

"If the stimulation in this part of the brain is continued long enough, it will sometimes actually result in climax for the woman. For the man, the stimulation itself is extremely pleasurable."

When you stimulate subjects in the right way, research analysts report both physiological and psychological arousal . . . physiological by showing the sex flush, the change of the color of the skin, erection in the man . . . all those things. There is also psychological arousal, usually in form of fantasy or memory. The person begins to remember sexual episodes of the past, or he begins to anticipate planned episodes for the future, and begins to fantasize about sex.

This discovery was made possible thanks to the cooperation of a bunch of delighted laboratory rats.

The rats were first rigged with electrodes for general brain stimulation. They seemed to enjoy the kick, and tripped the switches for their stimulation about twenty or thirty times each hour. Then the electrodes were implanted directly in the hypothalamus area of the brain, the area of sexual sensing. Guess what happened!

The little creatures just went crazy! They tripped the switches to get their pseudosex buzzes something like two thousand to four thousand times per hour and were so enchanted with the stimulation that they kept it up until they literally dropped from exhaustion. I think the expression "tired but happy" best describes them after their orgy in the interests of science. After they rested from their ordeal, they scurried right back to the switches for another workout.

194

The psychologist told me that the pleasurable effects of the electrodes will continue as long as the control box switch is turned to "on," and that the box switch can be operated either by the subject himself or by another person.

What a way to get your wires crossed! Here's one little gimmick that never has to be a turnoff, and since the device runs on batteries, not even a power failure can ruin your "impulses" unless you're so careless as to let your batteries run down!

I fully expect, some year soon, to find a sterling silver "his and hers" electrode set advertised in one of Neiman-Marcus' great Christmas catalogs!

If you think the sex machine is pretty far out, consider another new development that's helping couples enjoy the wonderful game of sex.

The penis implant, a plastic surgery procedure to give substance to the penis so that the man can obtain and maintain an erection, has been performed for about four years now.

Physiological or psychiatric problems may cause the man to be unable to satisfy his partner, but with the implant, the man is able to fulfill his role in sex.

The operation is done under general anesthesia and causes the patient little or no discomfort. Silicone, the same synthetic that is often used in plastic surgery reconstruction techniques, is inserted along the shaft of the penis. Hospitalization time is a mere two to four days.

The real beauty of the operation (aside from allowing the patient to be a much, much better lover) is that once it has been performed, there is no visible evidence. The happy male can go about his lovemaking, and only his plastic surgeon knows for sure!

Chapter 10
How To Wine And Dine A Date More Effectively!

When you invite someone out (whether your wife, mistress, new love interest, or what have you) and she accepts, it seems to me you have a moral obligation to have something planned for the evening. At least have a suggestion or two. To me, and to many of the women I talked to over the years, there's nothing so wishy-washy as a man who says, "What do YOU want to do tonight?" or "Where do YOU want to eat?" He may think he's being very gallant and thoughtful, but many women are turned off by such tactics. Instead of coming across as a generous, solicitous man, concerned over her preferences and welfare, he comes across as a man who lacks imagination and doesn't take command with authority and strength.

Psychologists tell us that women are irresistibly attracted to men who are competent in coping with situations, both in their work and in their social lives. This admiration and attraction to competence is very often placed above even physical characteristics. Can you just imagine Humphrey Bogart saying, "Anything special you'd like to do tonight, baby?" Or Aristotle Onassis saying, "What do you feel like eating tonight, Jackie—Jewish, Italian, or Chinese food?"

If you want to make a good impression on your date and

197

stand a better chance of scoring at the end of the evening, start concentrating on building up your positive image in her eyes.

Consider the following invitations and put yourselves in the girl's shoes for a moment:

"Where do you want to go tonight, Gladys? Are there any good movies playing downtown tonight?"

"Are you getting hungry?" "What would you like to drink?"

Now compare these "implied" invitations with the following:

"I'm going to take in the new Dustin Hoffman movie at the Village East! Want to join me?" (If she doesn't, she doesn't; but you've still come up with an *idea*, not just a blah approach.)

"I'm starved! Hey, let's stop off at that Arabian restaurant down in Market Square and get some of those marvelous meat rolls wrapped in fig leaves!"

"Listen, Shirley, why don't you try the Wahine! They float a gardenia in it and serve it in a coconut shell. I really think you'd dig it!"

Do you get my point? Always be as specific as possible with your suggestions. First, it shows that your brain cells are perking; second, it shows, indirectly, that you're thinking about *her!* (If she says she doesn't dig Dustin Hoffman, or fig leaves, or flowers floating in her drink, don't worry—she'll come up with an alternative, and you'll both be spared those uncomfortable silences that mark moments of indecision.)

Having definite ideas also shows that you get around . . . know what's going on . . . or at least read the papers or magazines. You couldn't have come up with those suggestions if you were living in limbo.

Whatever plans you make, take the masculine lead and make those plans with oomph and vigor. She'll love you for it.

Your dating life ought to reflect your uniqueness as an individual, your thoughtfulness as a person, your tenderness as a lover. You can't blame a gal for assuming that the man who so graciously went out of his way and took the time to plan and execute such a perfectly *divine* evening holds the promise of being a thoughtful, generous lover in bed!

If you get in the habit of asserting yourself with positive actions in planning relatively simple things like dates —who knows—you might also wind up being more commanding, more authoritative, more knowledgeable on the job and in the sack. It sure doesn't hurt to try!

Trish, who at twenty-six has risen to be "Mother Bunny" at a swinging Playboy Club, has the opportunity to watch all sorts of men in action. She sees them in groups with other men, on luncheon dates with their secretaries or female clients, and on dinner dates with their wives, mistresses, or whomever else they are trying to impress.

"There is such a thing as 'class' in men," Trish says. "But the sad part of it all is that so many men think that class is bought with money . . . lots of money.

"This simply isn't true. Class doesn't have anything to do with money, but it does take time and the willingness to acquire the know-how. It's a matter of conduct and bearing . . . watching others with gracious manners . . . and imitating them. It's all tied in with learning, growing, but above all else, *wanting* to please others by your deeds and gestures."

I asked Trish about the mistakes men make in public that turn women off. She listed them and commented as follows:

1 Drinking to the point of intoxication. Perhaps men do this to show how rugged they are, what great capacities they have, but in effect they are showing their lack of confidence, their nervousness. They are admitting that they are ill at ease in this situation and need to bolster their egos with booze.

2 Loud and boisterous conduct. This is most apt to happen when men are in a group of men. They

try to exhibit one-upmanship and superiority with
the opposite sex by getting overly familiar with
the bunny serving them or by attracting attention
by raising their voices or telling ribald stories.

3   Complaining loudly about the food or service (or
lack of it). If the complaint is legitimate, management certainly wants to know about it, but
there's a way to do it without creating a scene.
Some men think it marks them as connoisseurs
if they make their protestations in such a way that
people across the room can hear them. The funny
part of it is that this type often makes an ass of
himself in the process. I remember one man who
very dramatically summoned me to his table and
in a voice raised several decibels over his normal
speaking tone complained, "This soup is cold!"
I must admit I got a great deal of satisfaction in
answering him, "It's SUPPOSED to be cold, sir.
You ordered Vichysoisse!"

4   Would you believe some men still use toothpicks
at the dinner table? I can think of nothing more
offensive, and many of my "girls" tell me this turns
them off, too.

5   Men who "fuss" unnecessarily. Few things are as
unnerving as a man who fidgets by slicking his
hair down constantly and keeps readjusting his
tie, pulling up his socks, or chain-smoking. These
signs are dead giveaways that a man is uptight
and lacks confidence in this situation.

"Any other tips?" I asked Trish, and she said, "Men
should remember to give their dates change when they go
to the ladies' room and it's the kind of a place where there
might be an attendant on duty. It's a gentlemanly thing
to do and women really appreciate it. It's a small gesture,
but a man who thinks of it is a big man. Getting her wrap
for her at the hatcheck counter, holding doors, etc., are
some of the more obvious niceties, but you'd be surprised
at how many men fail to observe them . . . even when

it's obvious that he's trying to impress the girl in other ways."

Other club and restaurant people I interviewed stressed that a novice at wining and dining out ought to work at knowing what it's all about. It doesn't hurt to be familiar with the menu so you can suggest the entrée for your date. It might even be a good idea to stop at the restaurant the day before the date, ask for the next night's menu, and study it or make inquiries so if she asks a question you'll have the answer.

"You'd be surprised at how helpful maître d's can be to a novice," said the proprietor of a well-known French restaurant on New York's west side. "Every good restaurant has a maître d' who makes it his business to know his customers by name after only a few visits." The best way to establish your name in the mind of the maître d' is to call ahead and reserve your table. Then, when you arrive at the dining spot, you identify yourself by name and say, "I have a reservation." After only a few trips your name should be indelibly etched on the maître d's mind. You'll be off to a proper start when he starts greeting you with, "Good evening, Mr. Bonvivant, so nice to see you again!"

The maître d' will be even more apt to remember you and your handsome face if you treat him properly. Depending upon the size of your city and the status of the restaurant, your tip might range from three to ten dollars. The first few times, you might tip as you *enter* the restaurant; when you're assured that your identity is known, tip upon leaving.

A famous star of Westerns has endeared himself to the maître d's of the restaurants he frequents by sending his gratuity to their homes with a little note, thanking them for their wonderful service. You can understand how thrilled and pleased the recipients are to get a personal note *and* tip at home, where their families can be made witness to the fact that they are admired and respected at their jobs . . . by none other than "Mr. Big"!

Overtipping, or tipping even 15 or 20 percent when the

201

service has been careless or below par, can be a sign of insecurity. Don't tip if your waiter has been slovenly or resentful. You'll be doing him a favor if you call him aside and explain why you don't think he deserves a tip. If he thinks serving others is beneath him, he should get out of the business. There's nothing demeaning about *any* job well done.

Some diners are so intimidated and insecure, however, that they reward even the most inferior service. The offending waiter can continue to thrive unless people report him to his superior, withhold his gratuity, or request not to be seated at his station.

If the service is good or even adequate, a safe rule of thumb would be from 15 to 20 percent of the check for your waiter and for a cocktail waitress who brings drinks to your table. Counter help is not tipped in some areas, but in most large cities they get almost as much as those who wait on tables. This is a personal thing, and some people are adamantly against it. I feel that if you have received good service with a smile, you should be willing to reward the one supplying it. Remember that their salaries are generally quite low and that most of them depend on tips to make a decent living.

Many restaurants, in case you plan to flash your credit card and add your tip to the food and beverage bill, are now asking that you tip in cash to avoid the time lapse and six percent off the top which most credit card companies take (yes, even on tips).

If you can't afford the posh dining scene, or if such spots make you feel ill at ease because you're afraid of doing the wrong thing, why not go the "camp" route?

One delightful young man I know makes it a point to find tiny, out-of-the-way places to take his dates. Some might call this slumming, but not Pete. Besides, the city where he lives doesn't exactly abound with smart little French restaurants.

"I have made a study of the most elegant tacorito and enchilada houses in San Antonio," he'll announce. His date will expect him to name one of the Alamo city's famed riverside Mexican restaurants. Instead, he'll take her on an expedition and uncover tiny, almost hidden restaurants in which Mamacita welcomes him with open arms and retreats to the kitchen to produce the best Mexican food this side of the Rio Grande.

Pete's date will spend the whole week telling her friends about the fun time she had—and Pete's slender budget isn't set back more than eight or ten dollars (including drinks at his apartment before or after dinner) for a really different evening.

Another young man I know, who's more resourceful than he is completely honest, has worked up a trick that serves him well, although I'm not sure anyone less imaginative could pull it off. This fellow will seek out a small, unpretentious restaurant that oozes atmosphere (hopefully family run). He will approach the owner, manager, or maître d' with the proper degree of deference and hesitancy and confide that he plans to bring his loved one there tomorrow night for dinner. "This will be a very special evening for me, you see, because I plan to propose to my girl friend and I want everything to go *just* right!"

Most people are romantic at heart, whether they admit it or not, and my friend is assured that everything will be done exactly to his order and taste. Of course, the girl he arrives with will know nothing of his arrangements, but she can't help but be impressed and more than a trifle awed by the attention he commands. People snap to when he arrives with her. They are ushered to the quietest, most intimate table for two. Invariably there are flowers on the table, even if there are none on any of the other tables. A waiter will be at their beck and call, even to the point of ignoring other diners. The help in the place will give them sly, knowing smiles . . . and keep looking their way expectantly. After dessert and over coffee and an after-dinner drink my friend will take his date's hand in his, look into her eyes deeply, and in his most dramatic, romantic voice whisper into her ear, "May I kiss you as a perfectly delightful end to a perfect meal?" The girl may blush, but she'll almost certainly nod her approval. At which point our hero will not simply peck her cheek but draw her closer, and with everyone in the restaurant beaming approval, kiss her full on the lips. Sometimes a cheer goes up as the help guesses she's accepted his proposal, but more often than not a bottle of champagne will be sent over for this "special event" . . . compliments of the house!

I'm sure the girls who have gotten this treatment still can't figure out what special charisma this young man has that makes people fawn and swoon all over him. At least twice my young friend was told that the check had been "taken care of" by either management or compliments of one of the other diners who wanted to know what "all the fuss" was about and why "that young couple over there" was getting all the attention and service.

Needless to say, my friend could never pull this trick twice in the same restaurant. In fact, he rarely returns to the scene of the crime . . . and little wonder. But I can't help but be curious as to how often he's made out because of the "unforgettable" evening he's provided his date . . . at relatively small expense.

You can almost always flatter a woman's motherly instincts with the "come help me cook" gambit. Stewart, an intern at a southern hospital, was fond of announcing he had a "care package" of frozen lobster and mussels airfreighted from friends in New England. Would his date please come over and help him decipher the cooking instructions? Often, shrewd Stewart would include another couple in the party.

A grand evening was usually had by all, with Stewart's date showing off her expertise at the range, Stewart providing the staples, apartment, and soft music, and the other couple volunteering to bring the wine.

Stewart used this ruse with dozens of girls until he finally married, and no one ever guessed that "helpless" Stewart had actually purchased the seafood at the local fishmarket!

If you've got scruples and these devious means of impressing your date are beneath you, at least take her to the best Japanese or Chinese restaurant you can find. When you arrive, insist on taking off your shoes and play footsie under the low tables. Second, insist on chopsticks with dinner. (Old Chinese proverb say chopsticks invented *after* fork but invented by very wise man so diners would have hours and hours to feed mind with good conversation.)

By the time you've floated a gardenia in her Wahine and played footsie over the tempura, she should be on her way toward thinking you're a pretty different type of guy . . . hardly the type who'd stick by the dull old television set for that lifetime of weekend afternoons to come, hopefully.

I trust you're ready to concede that there isn't a causal relationship between the amount of money you spend on a date and the amount of fun you get in return. I don't believe one has very much to do with the other. If you like a person well enough to ask her out, you should have a great deal going for you if she accepts. You've got the *basis* of a fun evening, now all you need is a plan. The plan, of course, might take into consideration your finances, the time of day, the time of year, the amount of time she can give you, and her life-style.

If she's a high-fashion model, for instance, and can't stand a hair out of place, you wouldn't suggest a camping trip or a hayride. Perhaps a movie or a sightseeing trip in the city would be more apropos.

Here are a few suggestions, but I'm sure you can come up with some doozies if you just think about it:

1 Go to a drive-in movie with a box supper packed. If you've got a station wagon you can *really* spread out . . . in more ways than one.

2 Go on an evening picnic. Along with wine, cheese, and Italian or French bread, pack a candle. It may come in handy in the course of the night! Have it in a park, out in the country, or even in your own backyard or on your

rooftop . . . if it's flat, naturally. (Almost lost you there for a second.)

3 Rent horses, bicycles, motorcycles, snowmobiles, and make a day of it. If you happen to HAVE two of any of the above, you're halfway home. And think of the money you saved. Spend it on something else to make the outing memorable.

4 Discover a new, off-the-beaten-path restaurant. Form a group if you like, and each week or month have someone else in the group be responsible for coming up with a place no one else has ever been to.

5 If you live near the beach or a lake, use that as a setting for some of your dates. Some beaches have amusement parks, but better yet is making your own fun. Swim, hike, row, fish, dig for clams, but best of all, be on the lookout for secluded coves or deserted stretches of beach. They make ideal settings for just foolin' around.

6 Meet for lunch or breakfast instead of dinner, especially on days you both have off. It costs half as much as a rule, and you've got a lot of time ahead of you. Some people *need* all the time they can get, and others *like* all the time they can get.

7 Go sightseeing. Explore a specific part of your city or state. Take an excursion, visit a zoo. Watch the animals. You can learn from them. Take a moonlight sail up the river. Watch the other passengers. You can learn from them, too!

8 Pick a sport you both like and become participants, not just spectators. Ski, ice-skate, roller-skate, ANYTHING, but get moving. The exercise will stimulate you for who knows what?

9 Invite her over for breakfast in bed . . . or better yet make it a pajama party (and forget to mail the other invitations). Fix kippers and eggs or flapjacks. Sit Indian-fashion in the center of the bed for your repast. You'll be in

the right place at the right time if either of you
decides to take a siesta after such a magnificent
meal.

10   Have a meal catered . . . at her place or
yours. Plan the evening so there's a "theme."
If you send out for oriental food, place cushions
on the floor around a low table—remember
candles, soft music, incense . . . the works.
Make it a Mexican meal, or Italian, Spanish,
Arabic, whatever. Try to do it right! Stuff her
fortune cookie with a personalized message like
"Tonight's the night" or "Who loves you more
than I do?"

I'll bet you can come up with some entertaining ideas
that would make Perle Mesta look like a church social or-
ganizer, once you get into the swing of wining and dining
with flair.

Promise yourself never to settle for just an evening out.
Remember that women are absolutely dazzled by atmosphere—much more so than most men. (Remember what
I told you about making love . . . how women are much
more emotionally affected by soft lighting, by incense, by
different and romantic settings? The same is true of wining and dining. Any experienced hostess will tell you that
she could serve a TV dinner—on her beautiful china, ac-
companied by the proper wine in shining crystal goblets,
with the dining room lit by tall tapering candles—and get
raves for the meal!)

You must do the same with your wining and dining.
Create an atmosphere, set the scene, use your imagination
lavishly rather than just your pocketbook, and you'll make
an unforgettable impression as you wine and dine your
way toward a lasting and lovely relationship.

Chapter 11
How To Tell If She Really Loves You

Although you'd like to believe that every woman goes out with you because you're handsome, charming, witty, and sexy, the truth is that you may be filed away under a number of categories in women's little black books!

We've all heard of predatory females, and all men like to think that they can spot them a mile away. Some women, however, are so subtle that you may not even be aware you're being used to help her achieve a desired goal.

You may be the only millionaire in her life. But you may also be the only father figure, the only business adviser, the only doctor, the only lawyer, or her favorite teacher, electrician, plumber, or poet.

She is busy playing the game of life, and so are you.

You must understand that in this game there are many, many jobs to be done and that for every job there is a tool.

Her needs at the moment demand a certain tool for a certain job. Does she want her tax return completed with your new CPA stamp at the end of it? Or do the woofer and tweeter on her stereo demand your electrical genius?

Is she after your big brown eyes, or your love equipment, or your business acumen, or the whole, wonderful, complex person that is the entire, uniquely individual you?

In short, is it love that she feels for every part of your being, or is she merely using those parts of you that she feels may be useful at a certain point in her life?

From all my experiences in years of dealing with women, let me introduce you to the basic types of the so-called weaker sex any one of whom may fall completely in love with you . . . or who might happen to use you at one time or another (very often depending on the chance timing which guides when you come into her life!).

Here are the basic types of women you may encounter. One of them is probably sizing you up right now, so learn to recognize her!

FIRST, THE LEANER. This frail bit of femininity never outgrew her need for a "Daddy." He was the one who never let mean old Mommy spank his little darling. He pampered her with doll houses and trips to the zoo. He fought all her battles at school. She probably talks about Daddy quite a bit. Well, Daddy's retired to Palm Beach (with a mean old second wife) and our leaner is caught alone in the great big city, with no Daddy to protect her from the horror of it all (that nasty woman at the bank who keeps insisting she's overdrawn, the roommate who acts snippy when she leaves the apartment a mess, the landlord who complains because she wants to keep an itsy-bitsy poodle on the premises).

She's searching now for an everlasting Daddy, and will transfer all her weaknesses to your big strong shoulders if you don't watch out. If you want to take on her check balancing, her fights with roommates and landlords, in short, her battle against the frightening world, have it! She'll make you feel ever so strong and manly for a while. But guess what! Two children and ten years of marriage later, she'll still be calling you at the office every time a light bulb needs replacing, or the milkman leaves too much half-and-half, or little Tommy's nursery school teacher says he's been misbehaving again.

Then, somehow, you'll keep trying to remember why that cute, dear dependence of hers was so appealing—but the two of you will be stuck as "Daddy" and "Baby," with you wishing just once she would act her age!

SECOND, THE SELF-PROMOTER. You've got that special something that's going to help her forge her way upward to her next personal or career goal. She might be a model, a starlet, a copywriter, or a social climber.

You can figure quite lovingly in her plans until she's used you to overcome her present hurdle and is ready to move on to the next sucker.

And don't assume that because you aren't head of a talent-booking agency, you can't be used.

You might be chef at the top hash house in Paducah and find a young girl who wants to move up from salad arranger to fry cook—and will be happy to use you to promote herself to this new goal. This situation is simple to recognize: Simply assess your job or position in life and look at the women who are currently "beneath" you. Then play your cards accordingly.

THIRD, THE MARRIAGE-MINDED. Need I say more?

It might be your money, or it might be her childhood image of herself in any man's ivy-covered cottage. She might just love you madly, or she might just be in search of anything in pants (the kind that zip up the front for a specific purpose) just to save her from spinsterhood.

This type, if I were writing ten years from now, might be about as extinct as the whooping crane because "women's lib" is making ever so many women less willing to settle for any man or any marriage that doesn't offer them better rewards than their careers!

But right now there's a predatory breed that has one goal in mind: seeing a replica of you etched out in icing atop her wedding cake. Most of this breed of women are quite happy to "play married" before the rites are said . . . but unless you make your permanent bachelorhood clear, you're likely to find yourself irrevocably hitched when all you had in mind was a fun, carefree fling!

FOURTH, THE UNHAPPILY MARRIED WOMAN.

Ouch! This is bad news. You can meet her almost anywhere, but I'd place my bet on your meeting her at a
cocktail party at which her husband, "poor old Harry," is
also present. She won't hesitate to tell you what a trial
Harry has been to her through the years and how unhappy
her lot has been.

You can fall for this as easily as so many women have
fallen for men's my-wife-doesn't-understand-me gambit.

But if you think about it for a minute, you'll realize how
weak this woman must be if she continues to live off poor
old Harry—and has never gotten up the gumption to divorce him.

If you get serious about her, she'll either refuse to divorce Harry—or, worse, may do so and marry you. Then,
in a few years, guess who'll be the new "poor old Harry"
she discusses with friends and strangers alike?

Do you really need to get involved with the unhappily
married woman?

FIFTH, THE VIRGIN. Oh, this category is very tricky.
We'll have to delve into several types of "virgins" to give
you a thorough understanding. Let's begin the virginal
ABC's.

A. *The Real Virgin.* For very beautiful moral or religious reasons, this girl simply isn't going to relinquish her
virginity before marriage. You can either capitulate and
meet her at the altar or seek your fun elsewhere. A sub-
category (I could call it A–1, but how unfair it would be
to widows and divorcées everywhere) is the girl who has
been married but has lost her husband through death or
divorce, and thus becomes a recycled VIRGIN. She will
not play the game of sex again until remarriage. She is
NOT, as so many seem to assume, an A–1 candidate for a
saucy and immediate extramarital affair.

B. *The Ice Queen.* This is a virgin who has stayed virginal only because she's as frigid as a fjord in February,
although she'd never admit it. She much prefers to place
great emphasis on her virginity. Many a man has felt so
fortunate to have found such a rare Arctic bird. He mar-

ries her, often voicing his pity for the poor guys who settled for slightly more used merchandise. There's only one flaw in his selection: six years of marriage later, he comes to his senses and realizes that his wife is still (if you could measure her actual participation in their marital fun) a virgin, and that the move from a double to twin beds will soon be followed by a move to separate rooms; he simply married the coldest female alive, and he can't figure out what he ever saw in her. (In the meantime, his buddies who married normal, hot-blooded American girls seem to be having the time of their lives!)

Psychologists like to tell us there is no such thing as "pure motivation," which is their nice way of saying that most of us go around doing a lot of things for reasons we really don't understand!

You'll note that so many of the women I've talked about seem to want to "use" you one way or another. How do you separate their love for your big brown eyes from the fact that they need you to help them in hundreds of different situations?

First, let's not think of the word "use" as being entirely bad. We all use each other to some extent, you know.

Second, remember that a girl might be really-truly 100 percent in love with you and *still* use you for any number of things. She can use you as a meal ticket, as a handsome decoration to take to parties, as a hauler of furniture when she moves, as a stepping-stone to a better job, as her ticket to getting things wholesale, or as a wonderful companion to keep her company for life.

"I have never minded the word 'use,'" said *Cosmopolitan's* vivacious and outspoken editor, Helen Gurley Brown. "And I wouldn't say that being used is the very worst thing that can happen to a man.

"Let's talk about how women use men badly first," she said. "There are women who are really very rotten to their men, and most of these women are *married* to the poor man. I sometimes marvel at what a husband will put up with in terms of the wife's refusing to sleep with him any

213

longer. Instead of throwing the silly broad out on her backside, the guy kind of hangs around, hoping she'll come back some time.

"He may have affairs; he may have a girl; but instead of repudiating the silly wife for not loving him, he continues to let her live under the same roof, use his charge account, have country club privileges, entertain—and, when the poor guy finally keels over and drops dead, that silly wife gets the house, the cars, the bank accounts, and the life insurance!"

Helen Gurley Brown said that even if the wife had been a good hostess, a good mother, it's still a "rotten thing" if she refuses her husband sexually.

"So," she concluded, "I would say that men are used not just occasionally, but quite frequently, by women who marry for material gain. This is not so terrible if the woman *gives* in return. A lot of girls who marry very wealthy men seem to become terrific wives. They give, and they do their part. So I wouldn't say that such cases are the worst thing that can happen to the man. If the wife is a good hostess and is faithful and doesn't play around, and if she likes the man and is adoring and knows how to wear clothes well and please him in bed, I don't think that this type of wife is 'using' a man. It's a matter of her doing her share, and keeping her part of the bargain."

I also asked Helen Gurley Brown about girls who marry rich men and then divorce them and expect huge settlements or excessive alimony.

She replied, "I think this is one of the worst ways of using a man. A girl marries a rich guy for his money, lives with him for a year or two, and then divorces him and wants to live the same way she did before. In other words, she's not going to have to put up with old Pete any longer, but she wants old Pete's life-style. What has she done to deserve it?"

She confided to me about a currently publicized divorce action.

"The wife is asking seven thousand dollars per week

from a well-known celebrity. Well, he may have been a bastard to live with, but I think as long as she can hang onto him and be reasonably happy and put up with whatever she has to, she deserves the life-style his money can bring. But, having divorced him, what has she done to deserve his income?"

If even being approached for seven-thousand-dollars-per-week alimony makes you think you can avoid being "used" merely by being poor, you're absolutely wrong!

You're usable, and you had better be quite aware of it. A fat bank account certainly isn't the only thing that makes a male usable!

You're usable:

1. *If you wield enough power or prestige in your profession.* Such power can range from being chief casting director for a movie company to being head resident in a hospital in which a bevy of nurses are all busily engaged in working their way toward the top. Yes, and you can even be the head hashslinger in a lunchroom and discover that the waitresses are "using you" for better shifts! Any time at all that you have people working for you, or are in a position to hire and fire, you're in a prime position for being "used." (And the sweet young thing who's using you may not even be aware that her motives are anything less than 100 percent pure!)

If you're smart, you can avoid being used. But even the most worldly wise men are sometimes put in a position of being used. A movie critic friend of mine told me this story. He's one smart cookie, a man of great perception, and, as you'll see, great modesty.

Having previewed a superbudget Hollywood extravaganza which was due to premiere in New York several weeks later, he was surprised to see the ingenue starlet descending in the elevator of his apartment hotel one morning.

Without identifying himself, or the movie soon to premiere, he said a polite "good morning" and mentioned that he was a great admirer of hers.

Little Miss Haughty snubbed him.

Several days later he merely nodded pleasantly enough as they again met while departing from his building.

The next day he was standing in the lobby talking to a big producer who recognized Miss Haughty and called her over. "I want you to meet ——, film critic of —— magazine," said the producer.

Miss Haughty did a double take, then spent the next five minutes gushing to my friend about his marvelous reviews, how she never missed his columns, and how she so hoped they could get together for dinner.

"I was cordial enough," my friend conceded. "But I certainly wasn't as warm as I was that first day in the elevator. You see, when I was just a plain human being, she couldn't afford to even return my conversation. When I became a critic who can help or hurt her, I rate a big gush. I was just glad I started out with her as Joe Blow—it showed me her true colors."

2. *If you're a lot older (and perhaps much less physically attractive) than the woman you date.* Oh, you can be a lot older than she is and be loved for your very own self, not just for the fact that your experience and your bank account make men her own age seem like Little Leaguers! But no matter how you think she feels, can you step away from the situation and take a good hard look at the two of you—just as your friends do?

If you're a Cary Grant, a Bob Cummings, a Bob Hope, a Dean Martin, a Frank Sinatra, congratulations! You've kept your good looks, your physiques, your general health, your physical condition so well maintained that you can compete for the attention of women young enough to be your granddaughters!

But if you really can't rate yourself as physically attractive and find yourself pursued by a much younger woman, you can bet your friends are whispering that you're being taken.

The trouble with a much younger woman, unless she's been around quite a bit, is that she's terribly likely to be overimpressed by your savoir faire in wining, dining, and lovemaking. You can consider yourself being used if she seems more interested in where you're taking her to dinner than in that nasty pain in your slightly arthritic tennis elbow!

Then there's even a worse complication. She'll go along with you for the fun and the new games. But, know what? Those men her age are learning all the ways to play the game of life, too, and in a few years they'll be ready to compete with you on every level (without the arthritic pains to bog down their passion).

3. *If she loves to take you home to Mommy and Daddy —but nowhere else.* I know a lot of women who have a "Mr. Clean" stashed around as sort of an "Exhibit A" for family holidays. You may get to consume a lot of home cooking, of course, if that's all you have in mind. But you might mull over the fact that you're most likely her Mr. Clean, her smoke screen for some wilder affairs she's having in the big city. In the meantime, her parents think Suzie's so safe out on her own in the wicked world because they assume all the men she sees are "just like that nice Bob whom Suzie brought home for Thanksgiving." (A lot of men have their Miss Clean, too, the rosy-checked all-American type who's great for family affairs but absolutely nothing when you consider starting a lifelong affair!)

4. *If she adores you to have you as her plumber's helper, her mover of furniture, and her provider of rides to work.* But when she wins a free trip to the Virgin Islands or goes to the office Christmas orgy, she usually asks someone else. If you have ever been in this situation and haven't faced the fact that you're being used, you must return to "go" and stay there until you wise up.

5. *If ever your sex life with her is "iffy."* This is probably the biggest clue toward her true feelings. I know there are women who will never go to bed with you. But I also know that there are women who make little contracts, "If I sleep with you, will you ——?" Such a situa-

tion would make you a sugar daddy, and automatically makes the girl a—well, you don't want to get involved, unless you're looking for a lifetime of trouble. Getting attached to a woman who uses sex to bargain for what she wants is about as cuddly as taking in a king cobra as a partner in sex. Of course, if you should ever marry one of this sort, you're signing up for a lifetime with an extortionist. This is being used in the worst way!

By now, I hope that you understand the types of women who may try to use you, and the factors that make you usable. Let's go through some of my tests to help you know if she really loves you—now that you're pretty sure that her "using" of you is just an innocent thing. Here are my tests for true love!

TEST ONE: *IF SHE REALLY LOVES YOU, SHE'LL WANT TO SPEND EVERY WAKING HOUR (AND MOST SLEEPING HOURS) WITH YOU.* A woman in love puts you first, and you know it. She never begs off from a date for anything other than an emergency appendectomy. She never leaves to visit cousins in Oshkosh, unless you consent to go along. If her schedule is crowded (every woman will have to wash her hair, or bring work home from the office, or grade papers some evenings), she'll ask you over anyway and let you watch television or play solitaire while she finishes her necessary work. She wants you around, and she lets you know it!

TEST TWO: *IF SHE REALLY LOVES YOU, SHE WANTS TO KNOW ALL ABOUT YOU AND FINDS THINGS ABOUT YOU FASCINATING.* A woman who loves you will sit by the hour and exclaim with your mother over your baby pictures. She'll be intrigued by all your boyhood accomplishments—flip over the fact that you were the youngest ever to make Eagle Scout in Kickapoo County. She'll try to learn all about what you were doing and what sort of person you were in those years when she wasn't lucky enough to know you.

TEST THREE: *IF SHE REALLY LOVES YOU, SHE'LL BE CONSIDERATE OF YOUR FINANCES.*

A woman who loves you wants to be with you . . . and if it's the real thing, she'll be as happy wandering through the zoo as attending a ten-dollar-per-ticket play. She'll be as happy munching a hot dog as eating pâté de fois gras. She'll be as happy spending a lazy Sunday afternoon playing Monopoly as going to an expensive resort. (Please don't interpret this to mean you should be a tightwad. Be prudent with your money, have all the fun you can, but watch out for the woman who seems determined to bankrupt you. If you think it's bad when you're dating, can you imagine how it would be to have a big spender as a *wife?*)

TEST FOUR: *IF SHE REALLY LOVES YOU, SHE KNOWS YOUR LIKES AND DISLIKES.* Recently we entertained two other couples at dinner. Because Mark's wife of six years was out of the room when coffee was served, I asked Mark whether Linda used cream and sugar in her coffee. He looked taken aback. "Huh?" he asked. It turned out Mark had no idea if Linda used cream, or sugar, or *salt* in her coffee. And after six years of marriage? I frankly don't hold out much hope for the marriage if Mark's that ignorant of his wife's likes and dislikes. But back to you and your problems. A woman who really loves you will learn not only how you like your coffee but how you like your steak, what kind of salad dressing you like, that you're like a kid about chocolate sodas, and that you can't stand to have dinner until after eight in the evening. She cares about you so much that she mentally files every little like and dislike of yours. If you've told her shrimp gives you the hives, you'll never be invited over for dinner and find she's gone overboard with a shrimp curry.

TEST FIVE: *IF SHE REALLY LOVES YOU, SHE LOVES YOU IN SICKNESS AND IN HEALTH.* Maybe you're easy to love when you're your buoyant, cheerful, sexy self. Does she care as much when you're a big nasty, red-nosed brute, snuffling with the worst cold anyone ever had and with a roaring temperature of one-half of one degree? If she really loves you, she'll be the first one at your side with a steaming tureen of homemade chicken soup. If you're hospitalized, she'll be there *every* day, yes *every* day, to visit. She'll be as patient and as concerned as your mother ever was, and heaps more fun, too!

TEST SIX: *IF SHE REALLY LOVES YOU, SHE'LL CONFIDE HER INNERMOST WORRIES, AND SHE LIKES TO HAVE YOU CONFIDE IN HER.* A woman who really loves you wants you to understand her deepest feelings. She'll want to talk about everything, from her disappointments and little triumphs at the office to her views on the world situation to . . . yes . . . babies. If you really love her, you'll listen. I think that every couple who's really tuned in to each other can talk for hours and hours and hours, and still find plenty to say to each other. You do each other the courtesy of listening, and listening well. You provide for each other that reassuring ear that means so much. It's a wonderful feeling to think, "Why, she feels the way I do" or "She needs love and under-standing just as much as I do."

TEST SEVEN: *SHE LISTENS CAREFULLY, LIKES TO HAVE YOU GIVE HER ADVICE, AND FOLLOWS YOUR ADVICE WHENEVER POSSIBLE.* In one of the happiest marriages I know, she is an ex-model now teach-ing several university courses, and he is an accountant with a big oil firm. Sandy, the model turned teacher, re-spects Hank's opinions so much that she virtually goes on a fast if he suggests that she's put on a pound or two. A woman who really loves you respects your opinion so much that she accepts it, thinks about it, and tries to follow it through. She looks upon criticism as being given only be-cause you care enough about her to offer it in the first place. (And besides, you always make your criticism very, very gentle and loving, don't you?)

TEST EIGHT: *IF SHE REALLY LOVES YOU, SHE ANTICIPATES MOST OF YOUR NEEDS.* A woman in love is like a skillfully trained surgical nurse. She knows what you need *before* you ask for it. Ellen, a young house-wife with several small children, runs into her share of problems with child rearing, insufficient household help, and trying to keep up her considerable talent at painting. "Every woman has days when every single thing seems to go wrong," Ellen told me. "In fact," she continued, "I seem to have more of those days than other women do . . . maybe I'm just not as organized. But the day Bobby flushed four tropical fish down the john, then jammed down so much toilet paper that I had to get the plumber to unclog the toilet, I certainly didn't hit Phil with all the day's disasters when he walked in the door. I can tell when he's had trouble at the office or when he's blue and discouraged. My time can wait . . . when I can see he needs a decompression period from his rat race at work, that's what he gets!"

Anticipating your needs is simply thinking of how to please you—without being asked. Another wonderful young wife I know confided in me that their marriage actually became stronger during her husband's recent and lengthy hospital confinement following a car crash. "Bill was getting grouchier and grouchier," said Janet. "I knew what was wrong, too. Thank goodness he had a private room because one morning I just closed the door and gave him oral sex. He didn't have to ask me; I knew what was wrong. He had a need, and I knew how to take care of that need. It made me feel great to be able to give to him . . . and it meant so much to him because he knew it was a gift of love. After that first morning I simply brought one of those Do Not Disturb signs from home, and the rest of Bill's hospital stay was much more pleasant."

TEST NINE: *IF SHE REALLY LOVES YOU, SHE WILL BE ALWAYS SENSITIVE TO YOUR SEXUAL NEEDS.* She will want to please you every time you play, and how often will you play? Well, first let me warn you that the good Lord simply hasn't made a woman yet (as far as we all know) who isn't going to have some days and nights when she isn't up to par. A woman who really loves you isn't going to plead off with a migraine headache four nights out of seven. But you can expect at least several days a month when for one reason or another she won't feel like making love. Romeo, this is when you can improve your relationship even more—and say, "Baby, I want to wait until it can be really good for you." Mutual consideration is what counts. If she really loves you, she'll want you, want you, want you. She'll work to be at her best when you play together. She'll do her Kegal exercises to keep the tissues of her vagina resilient and firm so that she can give you more pleasure each time you make love. (And this is especially important if she'd had a child or two.)

TEST TEN: *HER EYES WILL LIGHT UP WHEN SHE SEES YOU.* The other night, at a political rally, of all places, I saw what I think is the most important indication of love. A woman I know, married about twelve years or so, was sitting at the rally with her young son. Suddenly, her husband walked into the room (I later learned he had been out of town and had not expected to make it back in time for the rally). The look on Carolyn's face when she saw Steve was everything that real love is. The look was glowing, tender, lustrous, possessing—all these things. There, I thought, is a woman who really adores that man of hers. Sadly enough, how rarely we see such a look! Next time you're in a restaurant, survey the couples you see. Half of them probably won't even be talking; what has happened to their love! Another fraction will be talking, but from the expressions on their faces, they must be discussing whether they can pay Junior's orthodontist bill before the family homestead in Ugly Acres needs a new roof. Then, with luck, you'll see another fraction of diners—who are broadcasting that look of love so strongly that you just know what their plans are as soon as they finish dinner! That's the look she'll be giving you if she really loves you, and don't ever forget it. If she really loves you, the eyes will have it!

I hope that your ladylove passed all the tests with flying colors, that you're now certain her love for you is strong, pure, beautiful, and that your future together can be nothing but golden. Yet there is a lingering doubt in your mind. . . .

Before you take the plunge into the sea of matrimony (or, if you're already married and worry about the big rocks in that blissful sea), take the time to find out what sort of family your love grew up in.

Was she happy and secure as a child? Does she speak of her parents with fondness? Have you met her parents and observed firsthand how she feels toward them?

Sadly enough, many women, because of their childhood and adolescent relationships with their parents, never became equipped to *give* love to a man.

The classic case is the girl who, for one reason or another, grew up hating her father. Perhaps he mistreated her or ignored her. Perhaps her mother disliked the father and constantly criticized him before the girl. Worse still, perhaps there was a nasty divorce and the mother spent years "sharing" her misery with the child, warning her that "all men are bad; never trust one!" This is a sad state of affairs, of course, but the things we hear in childhood, especially if they're the source of great unhappiness to a parent we love, make very great impressions on our developing personalities.

I've known many a young woman who is still terrified of sex because her mother, in giving her the facts of life, made her believe that she'd lose her virginity if she didn't watch for a big bad rapist who lurked behind every bush, waiting to attack her.

Perhaps you're satisfied that the woman you love has had a normal, happy upbringing. But if she seems to hold out secret reserves of herself and you sense her unwillingness to give herself up to you totally, check into her background! Sometimes it takes years of counseling to straighten out fears and hatreds acquired in childhood, so be forewarned if you've happened to fall in love with one of these unfortunate women.

Being unloved as a child or being led to mistrust men,

all feelings of tenderness toward you won't be enough. It may take years for her to learn to give herself completely, and trustingly, to any other person. If this is her problem, you've either got to have the patience or the psychiatric fees, or both, to help her over her problems. Otherwise, you're going to find yourself spending the rest of your life wondering why "she never seems to want me the way other wives want their husbands."

Chapter 12
How To Get In—And How To Stay In

I know you can learn from what some famous and really unforgettable men have told me about winning over a woman . . . and keeping her won over.

These men have mastered every trick of grooming, of putting their best foot forward in public and their best performance forward in private. You know how important all this is, don't you?

But these men have mastered something that is just as important: how to get in and stay in (her mind of course).

Physical attraction is lovely, but marriages and lasting relationships are played out in the living room as well as the bedroom. Your passion in a bachelor pad one year can turn into boredom in a bungalow tomorrow unless you work at keeping those sparks flying. She'll meet you more than halfway if you do, and, believe me, you'll have a relationship that will never go stale.

Have you ever heard a man ask, "What can she possibly see in him?" as he watches his love go off with her new lover? What she "sees" in him could be any one of a number of things . . . more than wealth, more than social standing, but in qualities that are more than skin deep . . . things that make him a superpersonality to be around!

I've interviewed movie stars and suburban matrons, college co-eds and call girls, and asked them about the men who made the biggest impression on them . . . men they couldn't get out of their minds.

Do you want to tune in on some of these interviews? You can steal any or all of these tricks, you know, because there's no patent pending on kind and thoughtful acts!

THE BIRTHDAY SWITCH

"The nicest thing I ever had happen to me was done by the most thoughtful man in the world," a well-known woman entertainer told me. "It was my birthday and he sent flowers to my *parents*, along with a note thanking them for bringing me into the world so he could meet me! You can imagine how wonderful they felt, and as for me . . . I just floated. It was the sweetest thing anyone has ever done for me, and I'll never forget it. He could have given me anything he wanted to, but those dozen roses he sent to my parents as my birthday present I'll never, never forget."

GOING PUBLIC

"A man who thinks about it," said a twenty-two-year-old college senior, "ought to know how nice it makes a girl feel if he *openly* says he loves her. I've had any number of liars tell me they love me *in private*. I think a guy who wants to let a girl know he's proud of her, and that he *really* loves her, will proudly tell his friends and her friends that he is in love. It makes a girl feel a lot more precious, and secure."

DIALING LOVE

"A husband who wants to have his wife stay in love shouldn't walk out the door to his office and just forget

her," a young wife told me. "My husband's first job, to help
support us when we were first married and both finishing
school, was in one of those awful telephone-soliciting
sweatshops. I was stuck at home with the baby when I
wasn't in class. It was awfully lonely for me. Ron couldn't
call home, of course, but he got in the habit of calling and
letting the phone ring just once before he went back to the
phone book page he was covering and made all his other
calls. It just let me know I wasn't *really* alone—and that he
was thinking of me and loving me."

PUPPY LOVE

"Can you believe I married Phil because of a dachshund
dog?" Nancy laughed. "I kept trying to tell myself I didn't
love Phil, but he was always doing things that I just
couldn't forget, and I'd find myself sitting in the office and
laughing over something he had done or said. Well, one
day he arrived at my apartment—I had already told him I
had another date at seven thirty—at about six, carrying
this silly-looking little Snoopy-type dog. The dog had a
note attached to his collar that said, "I need love." The dog
promptly licked me with a sloppy kiss then curled up on
the pillow and went to sleep. I think the gift of something
living, something that you can talk about together and
laugh about together, is one of the dearest things a man
can give a girl. I broke my date with the other guy that
night, and Phil and I were engaged two weeks later."

DELAYED ACTION

"Some men do things you never forget, and they seem to
be tuned in to what a woman really needs at a particular
time," a stewardess told me. "Once I was invited by a
passenger, a movie director, to a New York premiere and
a great party afterward. This fellow was so smart. He must
have sensed I was nervous about looking just right and

worried about what I was wearing. I mean, I dress as well as I can, but on my salary it's awfully hard to compete with movie stars!

"Well, he just smiled pleasantly when he picked me up and didn't say much. He waited until after we had walked into the party, then he leaned over and whispered in my ear, 'Karen, you're the most beautiful woman in this room.' His telling me then, when I was surrounded by all those beautiful rich women, did a million times more for my ego than it would have if he'd said it at the door to my apartment. What a sensitive person he is!"

CHECKMATE

"I think a woman has to know that a man feels protectively toward her," said Joanne, a twenty-six-year-old nurse. "The fellow I'm dating now lives twenty miles away and is finishing graduate school. It's hard for us to see each other except on weekends. But he knows I work the three-to-eleven shift, and a few weeks ago he started calling my apartment every night at 11:40 or so, just to make sure I got home all right. He *cares*, and I'm touched by it. When I told him the toll calls would add up and cost too much, he just laughed and said it was a pretty small price to pay to know I was home safe. And to think I used to date guys who would let me drive home alone from a party at 3:00 A.M., and never even call the next day to see if I was alive!"

HAPPY THURSDAY!

"I'm a sucker for men who give thoughtful little presents," a woman executive told me. "This may sound silly, but I almost resent the 'have-to' presents on Christmas and my birthday and Valentine's. Some men think they're doing just great because like a bunch of sheep they do what the retail ads tell them to and remember their wives or lovers on such days.

"I fall for men who send me a single rose with a note that says 'Because it's Thursday, and I love you' or one of those little poetry books inscribed 'Thank you for last night, Darling.' This shows real thought, and I'd rather have something like this than a fur coat every Christmas for the rest of my life." *

PZAZZY PLANNING

"I think a woman can pretty well tell during a dating period what sort of person a man will be for the rest of his life," one woman told me. "I adored dating Bobby because things always perked up when he was around. We never just went out on a date; it was always an exciting and off-beat thing. Oh, once he borrowed a friend's canoe and we paddled all the way down a bayou from a forest preserve area into downtown Houston. We must have looked pretty silly, but the laughs we had were worth the stares. Bobby's mind seems to work overtime at dreaming up different and fun things to do, and he certainly hasn't let a few years of marriage slow him down. A couple of years ago he pulled up in front of the house with an enormous camper—the kind that sleeps six and has air conditioning and a shower—that he had rented. About that time the maid who sometimes worked for us drove up. It turned out Bobby had decided to take us all on a two-week camping tour through the West, and the maid was to help me get the children's things ready and then come along to help with all the picnics and stay with the kids so Bobby and I could get out at night. Now, I ask you, how many women

* Author's note: While in perfect agreement with this woman's general philosophy, the author might readily be more moved by the gift of an Emba mink on Christmas Day then she would by a book of the mushiest love poetry inscribed "To Debbie . . . just because it's Groundhog Day . . . and I love you."

get a deal like that? All my friends have been dragged camping, and they always end up spending all their time washing dishes in a hot, buggy tent! Bobby knows I like to take the children places with us, but that I need a break in daily routine, too."

Are you getting the idea? Of course, you would have to know your woman before you could duplicate any of these gambits to stay in her mind. A girl with a severe allergy, for instance, might faint dead away at the sight of a puppy you brought her as a gift! There are millions of women who would faint dead away (with joy) if they found a mink coat under their Christmas tree.

But these general ideas, with your own personal refinements, can hardly go wrong when you want to be the sort of person she just can't get out of her mind.

1 Call her. Call her three minutes after you walk back in from a date, and say, "I love you; you're wonderful."

2 Call her to awaken her in the morning, and wish her a happy day at work. (This is, of course, assuming you're not sharing the same bed.)

3 Call her when you're out of town. Entertainer Kaye Stevens told me she has a wonderful Australian friend who will call her from anywhere in the world, just to say, "How are you? I'm thinking of you." Then he hangs up . . . mission accomplished. (Saves on his phone bill, too, to keep those conversations brief. But it makes him rate with Kaye because his calls are so frequent.)

4 Write her. Write her little notes to tell her you love her.

5 Write her when you're out of town. Nothing elaborate. A note dashed on a picture postcard of the Golden Gate Bridge, with your love and the notation that you're buying the bridge as a small token of your undying love.

6 Write poetry, if you can. If not, you might do

what a friend of mine did and compose zany,
passionate, love letters under the pseudonym
"J. Quinby Greebled." His wife saved every one
of the letters and said she married him simply
because she couldn't stop thinking about what
a lovable nut he was!

7   Write little notes just to be communicating with
her. Women love tangible evidence of a man's
concern. I know of a romance that came to full
flower simply because the fellow left a note on
her car windshield one morning as he left for
work. (Their parking spaces adjoined.)

8   Show her you're proud to let others know how
much you care about her. Introduce her to your
friends proudly, and make a fuss over her in
public.

9   Send flowers, but not just to her house. I
remember a girl who used to get a dozen roses
from her boyfriend every Monday morning—at
her office. She was pleased because he was
telling a lot of people how much he cared about
her.

10 Make your gifts so specialized that she'll know
you put a lot of thought into them. Give an art
lover a numbered lithograph (a one-of-a-kind
item, but relatively easy on your pocketbook).

11 Give a girl with a lot of blank wall space in her
apartment the tallest green plant you can find.
Your gift of something living, something she
can care for, is pretty special. (A woman I
know bought a small country place and was
moved to tears when the man she was dating
gave her two apple trees and had a nursery
plant them for her. "We may never see those
trees bear fruit," she said, "but every time I
look at them I think of Jim.")

12 Give her something that will bring the two of
you closer together. If she's admired your
ten-speed bike or your skiing gear or your
aqualung, why not make her next gift a real
surprise—and give a "hers" version of your
sports equipment? A big purchase? Yes, but

231

haven't we already decided that this is the
woman you're going to make the really
concentrated effort to impress so that you'll be
the *one* unforgettable man in her life and in
her thoughts?

13 Show her that you respect her interests. If she's
a ballet fan, take her at least once. If you
discover *Swan Lake* gives you a headache, you
can still show her you respect her love for the
art. Next time a performance is scheduled, buy
two tickets, beg off for yourself, and invite her
to take a girl friend along to the matinee. You're
saying, in effect, "Ballet isn't my cup of tea,
but I want you to enjoy it because I care about
you."

14 Give of yourself. If she's out of town, take her
widowed mother out to dinner. Your beautiful
gesture won't be forgotten for a long time, and
you'll get the girl's mother on your side, too!

15 If there's illness in her family, show your
concern. A man I know was the first to send
flowers to my father when he had a serious
operation. The man had never met my father,
but it was his way of showing what a considerate
sort of person he was. This was over ten years
ago, and I still haven't forgotten the flowers or
the fellow.

16 Be strong when she is rattled or upset. Never
write off things that upset her as insignificant
simply because they don't loom large in your
great big manly point of view. When she needs
help, give it, and show her the sort of person
you are in a crisis.

17 Show your everlasting concern for her health.
If she's starting to get a sniffle, gently steer her
home (even if it means missing the last quarter
of the hockey game) and tuck her gently in
bed, then offer to call her office for her, and by
all means call her during the next day; finally
(the pièce de résistance), bring her dinner that
night. Make her feel as though her sniffle is your
overwhelming concern and that the two of you

are going to whip the offending germ together. She'll love you for it, and long after the antihistamine has done its trick, she'll still have your kindness to remember with love.

You could have thought of each of these ideas, couldn't you?

Maybe a lot of these thoughts about giving have actually crossed your mind at one time or another. But did you carry them out? We say our lives are so rushed that we can't do this or we can't do that, and in the process we lose out on one of the most wonderful experiences we can ever have—giving unselfishly.

Giving, planning for someone else's enjoyment, and reaping all the satisfactions of making another person happy is such a happy experience for the one who gives that we can almost call giving a selfish experience. Why? In giving to another person, we experience a joyful feeling of doing for others and find our gifts returned tenfold. Giving somehow sets off a chain reaction of joy, so that everyone involved reaps the benefits!

By giving of your time and your consideration, you'll be repaid by getting in her every thought and staying in her mind forever—and by all the acts of love which she will give you in return.

Even if you never make it with her as a perfect and permanent duet in love, you can be the sort of person she'll *want* to remember the rest of her life. In making her life better and happier, even if it's just for a brief period of time, you'll make this world a little bit better. And isn't this what love is all about?

Chapter 13
How To End A Romance—
Without Getting Shot

In 1970 some 15,600 new tombstones went up in grave-yards across the United States, marking the final resting-place of lovers who didn't play fair at the game of love.

I don't know if any of the tombstones are inscribed "Here lies Joe; he done her wrong," but they all could be!

Have you mistreated women so badly in breaking off an affair that you count yourself lucky to be among the living? Or have you barely (no pun intended) escaped being shot, poisoned, stabbed, or run over by a woman bent on revenge?

Cheer up! Don't brood about the dear departed ones. You must consider that Joe, lying there with a lily in his hand, is at least at *rest*. I know a lot of men who are going through *living* hell today, simply because they hurt a woman they once loved.

You hear about the painful split-ups every day, be-tween single men and single women, married men and single women, and married men and married women. Oh, how things can get nasty when the breakup is botched!

You can break up with a woman and still emerge with your sanity, your health, and her self-respect intact.

But if you neglect her feelings, beware!

We hear about all sorts of breakups every day.

Beware as you read the stories of passion-turned-to-hate in the daily newspapers. (If you're smart, you can read between the lines. If you're rather naïve in what a botched romance can do to a man, read the translation at the end of each story as well.) Once you realize what dire fates await men who hurt women, you can protect yourself much better!

DENTIST IS DRILLED BY EX-WIFE

MONA LISTER REVEALS TAX EVASION TO IRS

Darien society dentist Dr. Donald Lister was convicted in Federal District Court today on eleven charges of tax fraud and evasion. The Internal Revenue Service charged that Dr. Lister, forty-two, had failed to report over $750,-000 in earnings since he began private practice in Darien nineteen years ago.

IRS spokesman Will Ketchum testified that the first clue of Dr. Lister's evasion of federal income tax came with receipt by IRS of an anonymous letter written on lavender-scented stationery.

A handwriting expert later identified the writing in the letter as that of Dr. Lister's ex-wife. Mrs. Lister testified in court that she "let the IRS know about Donald because he consistently two-timed me."

Mrs. Lister added, "He made a fool of me by running around with other women all those years that I thought he had Rotary meetings every Monday night, Kiwanis every Wednesday night, and Shrine every alternate Thursday. All this time he was seeing other women."

Dr. Lister testified that he became involved with a patient, Miss Kimberly Rogers, twenty-two, while doing a root canal. He denied charges by his ex-wife that he had "played with every single canal in Miss Rogers' body." He ignored his ex-wife's plea to the judge that he should "be sent up the river for playing outside his home canal." Dr. Dister denied charges of tax-evasion as well.

Further testimony from Dr. Lester's ex-wife, however, revealed that it had recently been discovered by her that Dr. Lister had established twenty-three savings accounts under assumed names.

Lister will serve two years in federal prison, and he faces indebtedness of over $310,000 in penalties, back taxes, and interest.

(Translation: Never overtax your wife's patience with your personal affairs, especially if you've underpaid your tax to Uncle Sam in your professional affairs.)

SPORTFUN, INC., EXEC IS STRIPPED OF ACCOUNT BY POWERFUL YOUNG WOMEN'S WEAR BUYER

Reliable sources have reported to *Clothing Daily* that Amalgamated Stores women's wear buyer Sharon Patterson has completely nixed the spring line of SportFun, Inc., creating the loss of that firm's major nationwide account (370 primary and suburban stores).

Miss Patterson, promoted to head buyer of Amalga-mated last February, was formerly with Goldman's of Atlanta, where she met Marty Feltman, executive vice-president of SportFun, and became a heavy buyer of his lines.

Clothing Daily's spies report that Miss Patterson was once on "very, very close terms with Marty Feltman" and that they appeared to be headed toward a "personal merger" until Feltman left town suddenly and alienated Miss Patterson by refusing to return her calls.

Their next meeting was in New York, after Miss Patterson had become head buyer for Amalgamated.

Miss Patterson reportedly told Feltman, after viewing the spring line, that "Amalgamated has already made its total commitments for spring." She refused to comment to this newspaper on her decision to drop the SportFun line. (Our spies report that Fashion Togs of California and Seventh Avenue's Fashion Modes will pick up the bulk of the orders formerly reserved for SportFun.)

With the loss of the Amalgamated spring order, SportFun, Inc., will reportedly operate some $2 million in the red and will close its Dallas factory.

From Los Angeles, Samuel Rothgerber, president of SportFun, stated that Mr. Feltman will step down immediately as vice-president of sales and that Robert Youngman, thirty-six, will succeed him.

Mr. Feltman has not announced his plans for the immediate future.

(Translation: Never misuse a woman for any reason. The little woman you hurt today may become the big woman who's libbed tomorrow and may sit in a mighty position, the better to trample upon you, your business, and your future.)

BUILDER'S FOUNDATIONS ROCKED BY EX-WIFE'S EXPOSE OF FINANCIAL DEALINGS

Pueblo Domestic Court Judge Homer Hartness today awarded what court sources call the largest monthly alimony and child support ever assessed in this county to Mrs. Lorena McMath Foreman, divorced wife of prominent builder Franklin G. Foreman.

Mrs. Foreman divorced Foreman three months ago, charging extreme mental cruelty and infidelity. She had been awarded temporary alimony of $500 per month and child support of $250 per month for each of the two minor children until domestic court could investigate Foreman's financial condition more thoroughly.

In announcing alimony of $1,000 per month to Mrs. Foreman and monthly child support in the amount of $750 to each of the couple's two minor children, Judge Hartness noted that the payments were based on a financial statement submitted fourteen months ago by Foreman, forty-nine, to three area banks. The statement was to substantiate his request for loans totaling $3 million for interim financing on his current apartment complex in northwest Pueblo.

Attorneys for Mrs. Lorena Foreman and the two minor children had produced the financial statement showing Foreman's net worth as of fourteen months ago as in excess of $650,000.

Foreman, who appeared highly upset at the introduction of the financial statement, testified that he was "unable to meet the payments as assessed by the court." He said, "I certainly do not mean to say I misrepresented my net worth to the banks in obtaining the loans for Foreman Construction Company. But my statement was dated over fourteen months ago, and things have a way of changing."

Foreman claimed to be unable to furnish an exact statement of net worth at this time but told reporters after the hearing, "I couldn't meet even one-fifth the payments that Judge Hartness awarded my ex-wife, and she knows it, too."

Foreman's attorneys immediately ushered Mr. Foreman from the courtroom lobby, where he met the new Mrs. Foreman, whom he married two months after his divorce from Mrs. Lorena Foreman. Mr. Foreman and the new Mrs. Foreman were seen in heated argument until lawyers hurried them into a waiting car.

Mrs. Lorena McMath Foreman and her attorneys emerged from the courtroom, apparently in a jubilant mood.

Earlier in 1972 Mrs. Lorena Foreman had been awarded full custody of the couple's two children, their home on Villareal Road, and their summer cottage in Conifer, Colorado.

A loan officer for First National Bank of Pueblo, who asked that his name not be used, told Pueblo *Press* today, "The FNBP is, quite naturally, investigating Mr. Foreman's financial status. We have no reason at this time to suspect fraud; however, the Loan Department is in the process of reassessing our negotiations to finance Mr. Foreman's Verde Vista condominium project."

(Translation: Never involve an honest wife in your dishonest business dealings unless you intend to be honest with her in your love dealings—forever!)

MEDICAL SOCIETY NEEDLES DOCTOR ON DRUG QUESTION

WOMAN IN WHITE PUNCTURES HIS "INNOCENT" PLEA

San Francisco neurologist Dr. Benjamin Kellerman has lost his license to practice because of misuse of drugs, the Marin County Medical Association determined in a four-hour closed session Friday morning.

Investigation into the case was begun by colleagues who believed Dr. Kellerman was "unfit to execute the practice of medicine" because of his dependence upon certain painkillers and sedatives now classified by the FDA as narcotics.

Sources within the medical tribunal told this reporter that the medical group was also disturbed about widespread rumors concerning the personal life of Dr. Kellerman. "The group felt his personal conduct was unseemly and was interfering with his practice of medicine," said a member of the tribunal.

(Rumors included Mrs. Kellerman's oft-repeated charge to acquaintances that her husband "scheduled the last appointment every day with the same woman." Some sources also reported that Dr. Kellerman's nurse, Mrs. Rachael Kean, had been romantically involved with the neurologist and became emotionally upset when Mrs. Kellerman called to ask about his conduct with the female patient.)

Both Mrs. Kellerman and Mrs. Kean were reported as testifying at the hearing. "Both testified that Dr. Kellerman had been relying increasingly on barbiturates to help him meet the pressures of his personal and professional life," a source told this reporter.

Although Dr. Kellerman may appeal his suspension to the State Board of Medical Examiners, inside sources predict he will not appeal the suspension and will voluntarily spend two or more years in a private institution before applying for renewal of his license.

The name of the third woman in the case, the "patient" who apparently triggered the testimony of Mrs. Kellerman and Mrs. Kean, was not revealed.

(Translation: Any man so square as to enlarge a love triangle to three women will be driven to drink or drugs. He will find an institution a very restful place in which to withdraw from all his bad habits.)

POLICE "SHOCKED BY SAVAGE MAULING" OF LOCALITE IN ATTACK AT HOME

"He looked like he'd been attacked by two lionesses," said an ambulance attendant who brought Anthony Brazell, thirty-two, of 1064 Hastings Place, to Mercy

Hospital Emergency Room at about 9:15 P.M. Saturday evening.

Brazell, a real estate salesman, was reportedly clawed, scratched, and beaten by his wife, Mrs. Mary Lou Brazell, and by Miss Georgia Dunn shortly after he pulled into the driveway of his home.

Partial excerpts from police files reveal that Miss Dunn and Mrs. Brazell told the police they discovered Brazell had been seeing both women.

"He told me he hated his wife, that she never understood him, and that he was planning to get a divorce as soon as he could," Miss Dunn told police.

Mrs. Brazell stated, "I knew Tony was keeping a mistress back in January. But he promised me that our marriage meant more to him than any other woman and that he would stop seeing the other woman immediately."

Police were told that when Mrs. Brazell was informed by a friend that her husband was continuing to see Miss Dunn, she called Miss Dunn and planned to "punish her."

Police said that the two women stated that when they met, they began to compare notes on Brazell's conduct, and both were enraged when they learned of his duplicity.

Brazell remains in critical condition at Mercy Hospital, and was unavailable for comment; Mercy Blood Bank officials are asking for donors of type AB blood for the victim.

(Translation: Early Christians in lions' den had somewhat better chance for survival than jet-age descendants who forsake certain basic commandments. See Exodus 20:14.)

After you do your Bible homework, you'll *certainly* know never to cheat on your wife.

Let's talk now about how you single men should break off with the women you date.

Although getting the woman you want and keeping her happy for as long as you want can be difficult, even more difficult is getting rid of her once you've outgrown her or want to move on to greener pastures.

You won't have any trouble knowing that you're falling out of love. You will begin to clock-watch when you're with her; you'll make excuses to leave her early. You'll stop remembering your little anniversaries, maybe even her birthday!

Unfortunately, most women don't want to recognize the signs that they're losing you. It's crushing to a woman's ego to learn that she has loved and lost.

Your method of break-off, then, must soothe her wounded ego and still get the final message across.

How can you let her down gently but firmly and leave her with the feeling that you are and can still be friends . . . and that breaking up is really to your mutual advantage?

Breaking up with someone you really care for as a person is going to require all the tact, gentleness, imagination, and sincerity you can muster.

"Honesty is the best policy" is a good rule . . . but not necessarily in this case. Sometimes the truth has to be dressed up just a bit to make it palatable. Sure, you can simply say, "I'm sorry, but I just don't love you anymore"! But if the girl is the sensitive, emotional type, think of how such a declaration might undermine her morale, her confidence in herself.

Wouldn't it be more gentlemanly, more gracious of you to say, "You're too good for me! I don't deserve you! I'd be a stone around your neck . . . weighing you down . . . preventing you from achieving your true potential."

What woman could argue with that? Oh, she might make mild protestations, but basically she'll have to agree, she *does* deserve better, she *is* too good for you! You have to be careful here to make her realize that as long as you're around she can't grow as a person or meet a more interesting, promising lover. You'll be very gallant and step out of the picture for *her* sake. You take all the blame . . . it's all your fault. You're a cad, rascal, no-good—the whole part and parcel.

If you can manage to break off in this manner, you've played the game to a fine ending, and you've left her ready to pick up the pieces and go on in life, with her head held high!

If, on the other hand, you treat a woman the way the man in the following story treated Sheila, you deserve to be thrown to the lionesses, stretched on the rack, or hanged by your thumbs!

Sheila's eyes flashed with spite as she told me of what turned out to be a one-night stand (a practice she's most unaccustomed to). The one-nighter, according to her sex partner, was only to have been the beginning of a beautiful, long-time relationship.

"Women are first turned on to sex in their minds," she said, "and when I met this guy on one of my New York–Denver flights, he just stimulated my mind so completely. I admired him, trusted him, and it was just great. You've heard the cliché 'love at first sight'? I thought it had happened to me! He made me feel it was mutual, so I let him pick me up. For the first time in a few years, I was really excited. I almost had an orgasm by just listening to him talk!

"We were lying on a couch in my apartment by candlelight. We both had our clothes on as I was lying on top of him. Before that it was just conversation. He knew I was leaving the country, and he said he wanted to see a lot of me before I went.

"We had just been talking and caressing each other for an hour. I was very relaxed, and very trusting. He may have tricked me, but it didn't seem to me that he was interested in making love.

"Oh, I felt he wanted to go to bed with me, but that wasn't the purpose of our evening. He tranquilized my mind . . . he was strong and masculine, but so sensitive, so witty.

"He had me hypnotized. When I first met him on the

244

plane earlier that evening, I thought there was no way I'd
go to bed with him. But he just mesmerized me . . . talking in a low voice and saying, 'Baby, "come" for me.'
There I was, fully clothed, being *talked* into orgasms!
And he wasn't even touching me in a passionate zone.

"Then I felt so bad because I was getting so much satisfaction . . . and we went to bed. By this time it was
about five in the morning, and he left at about eight. His
parting words were unbelievably cruel. He said, 'Well, I
knew your airline prides itself on their VIP treatment, but
I didn't know they were so dedicated that they attended
to their passengers' every need like this!' "

Sheila shuddered. "I never saw or heard from him
again," she said. "I felt so cheapened, so dirty. I trusted
him, believed him . . . oh, it was awful."

Sheila let her anger smolder for a few weeks before she
wrote the following letter. Since her playmate for the evening (the same one who had murmured to her about her
uniqueness, her mind, her body, and their future together) had not so much as given her his address, Sheila
had trouble mailing the letter.

Finally—and she feels not one twinge of guilt about this
—she simply sent it off in care of the fellow's boss!

Here's the letter:

Dear Customer:

In reviewing my books for July, I find that your
bill is still outstanding. According to my records,
you owe my firm for services rendered during a ten-
hour interval in mid-July of this year.

Since I have just recently "gone public" a
standard rate has not been established, but sources
advise me that similar transactions demand a sum
in the three-digit bracket.

To be honest, you were my very first customer
since starting into business and I was, unfortunately,
quite new in the role of prostitute. If you have any

suggestions for improved performance, please send them along with payment.

Of course, my legitimate employer is not aware of my new business venture, but I've considered bringing him in as a partner since his company provides my greatest source of contacts. You won't mind, then, being listed for tax purposes in such an event?

I wish to commend you on your fine salesmanship. My only suggestion being that you might leave your associate with a little greater degree of dignity at the termination of the transaction.

In closing, I wish to state that I have no doubt that your business endeavors will be fruitful, having witnessed first hand your skill in manipulating people and achieving your objectives. It is only wished that with such obvious ability one might strive to incorporate integrity as well.

Most sincerely,
Sheila Steward

If Sheila's story seems unusual, believe me, it's not!

It seems to me that Sheila's response was quite psychologically healthy. She got things off her chest (the memory of that lover and her rage) in a manner that helped her get over her intense hurt.

"KEEP A FRIEND"

Most young girls learn the song:

> Make new friends
> But keep the old
> One is silver,
> And the other gold.

Let's hope you don't take the "make" or "keep" too seriously!

My point is that you, as a man of eighteen, twenty-eight, thirty-eight, forty-eight, fifty-eight, and all the ages below, above, and in between, never know when you're going to need a real friend. This is the entire point of *never ending a relationship badly.*

If you manage the termination of a relationship well, you have created "gold" rather than "quicksilver." You'll emerge after the affair has ended feeling well equipped to live with yourself. The woman will feel like a *perfect lady*—not like Sheila who felt so cheapened that she felt she had to fire off that vindictive letter that used every bit of her brains and wit to fight against a lover-turned-adversary so she could rebuild her own painfully bruised ego!

If you're going to play the game of love like a pro, you certainly don't need enemies.

Babe Ruth and Lou Gehrig, Sam Snead and Joe Di-Maggio, Jack Kramer and Ken Rosewall, Wilt Chamberlain and Arnold Palmer, all these men are sports greats who beat almost all opponents at their own game.

Yet after the sports battle was finished, these superstars maintained the *respect* of those opponents. Why? Because they played the game with every sportsmanlike tactic—and never spit in their opponents' faces as they left the playing field!

Later, even after trades, reorganization of teams, or signing of new pro contracts, their former opponents were delighted to find such superstars playing on *their own* sides, simply because games won or lost, touchdowns, home runs, aces, and holes in one scored over them made no difference in the respect in which they held these players.

You should play the game of love the same way.

If it becomes necessary to break off a relationship, manage it in such a manner that your partner will remember your game of love as a highlight and remember you as a partner who never cared so much about "winning" that you had to emerge "top man" at any cost.

(Besides, you'll live a lot longer!)

Chapter 14
What Kills Love?

After you get an affair well started, or after marriage, it's only human to let your defenses down a bit. But if either partner thinks the game is won and decides he can rest forever on his laurels, your love is in danger!

As soon as you get sloppy about the things that mattered so much when you were dating, the spark of love will begin to die.

The gradual letting down of your good grooming, your considerate behavior, and your thoughtful techniques in bed are like the countdown to the end of her love for you.

If you're as careless as some men I have known, by the end of a few months of a close relationship love begins to wither on the vine. Your thoughtless omissions and commissions are killing love as surely as if you had planned to kill it from the very beginning!

If I had a referee's suit and a shrill whistle to blow and could call the fouls on American men, here's how I would deal out the penalties:

I'D PENALIZE A MAN FIVE YARDS FOR:

1   Appearing at breakfast unshaven or without a
     nice robe and slippers on.

2   Neglecting to brush his teeth first thing in the
    morning or last thing in the evening.

3   Failing to dress nicely for dinner. Can you
    believe that women still complain to me about
    their husbands running around in *undershirts?*
    As for appearing at dinner bare chested, I could
    condone it only if one's airconditioners had
    ceased functioning during a 108-degree heat
    wave, or if one were at a poolside barbecue, or
    a resort in Hawaii.

4   Failing to dress properly for the occasion. Even
    if your wife's old college roommate (whom
    you detest) is giving a party, you owe it to all
    concerned to dress as well as you would for your
    boss's most important event of the year. If you'd
    promise yourself always to dress for others as
    you would for your most important business
    deal, how happy everyone would be! And
    while we're on the subject, don't ask your wife if
    the brown suit with the tan knit shirt is all right
    for the Adams' party—and then get outraged
    when she says, "No, it's really not." You asked her
    opinion. You knew in the first place that the
    party was more formal—so don't take your anger
    out on her!

    Another unfair tactic is to ask her, "How does
    this tie look with my houndstooth jacket?" and
    then bristle when she tells you the truth—that it
    looks as though you combined the two in the
    dark recesses of your closet! When a woman gives
    an opinion, solicited or unsolicited, it's because
    she wants you to look your very best. When you
    ask her opinion, accept what she says gracefully.

5   Forgetting his manners with any houseguest.
    Even the oldest of friends should be treated as
    though they were visiting your home for the very
    first time. When Larry and Irma walk in the
    front door, even though you've spent ten years'
    worth of Saturday nights with them, you should
    try to be as great a host as you were on your
    first get-together. By that I mean you mind your

250

manners; you seat Irma at the dinner table;
you never let Larry fix his own drinks (unless
you're slaving over a flaming barbecue pit at the
moment); you make an effort at good
conversation. You play lord of your mortgaged
manor, and everyone will have a much better
evening!

6 Failing to back his wife up in her discipline of the
children. Oh, this is a bad situation. I think if
any marriage is going to survive the rearing of
offspring, you two great big adults have got to
present a great big united front! As a pediatrician
once told me, "Parents must operate on the basis
that the child is, of course, much smarter than
husband and wife put together." She was teasing
(I think), but her point was that adults must
set firm rules, with no loopholes. If children learn
that they can undo the discipline of one parent
by appealing to the other, they rule the roost, and
what a noisy, unsettled roost it will be!

7 Prefacing any criticism with either "you always"
or "you never." Example: "Marsha, you are
always losing your car keys." What you mean is
that she's misplaced her keys for the second time
in a month. No one *always* does the wrong thing,
and you're wrong to say so. The "you never"
accusation is similar. "Betsy, you *never* remember
to get my suits to the cleaners." Could this be
so? As you speak, you're standing there in a suit
that the cleaners pressed, and you're looking into
a closet with no fewer that five suits cleaned,
pressed, and ready to be worn. What you
mean is, "Betsy, you forgot to take my gray
double knit to the cleaners yesterday, and that's
what I really wanted to wear." The "always" and
"never" barbs are to be avoided. If a man tossed
such statements at me, I'd probably be so mad
that I *would* become the "always" or "never"
person he said I was.

8 Committing some of the minor, but nonetheless
annoying, sins of etiquette. You're going to kill

251

love if you eat like an eighteenth-century
farmhand at harvest time. Here's my quick list of
no-nos. Never bolt your food. (No one is going
to take your plate, so relax and eat slowly!) Never
park your knife or any other utensil with the
handle on the table and the business end on the
plate. Never, NEVER, replace any utensil, once
used, on the table—even if no plate or saucer is
provided. (Say, perhaps, you're confronted
with a custard cup of dessert, a spoon for the
dessert, but no plate to receive the spoon
when you've finished. Do *not* replace the spoon
on the table. Instead, leave the spoon in the cup.)

I'D PENALIZE A MAN HALF A FOOTBALL FIELD IF HE'S CARE-LESS ABOUT HIS SMOKING HABITS AFTER DINNER, EITHER WHEN DINING OUT OR AT HOME.

If you must smoke, it's more important than
ever now to ask the permission of your hostess.
"Do you mind if I smoke?" is a very necessary
courtesy because so many pollution-concerned
people *do* mind in this day and age. (You never
even ask permission while others are still eating.)
If permission is granted, you absolutely NEVER
NEVER NEVER use a plate as an ashtray. This is
absolutely revolting! If you're ever in doubt as to
whether a certain receptacle *is* an ashtray, be
sure to ask. Hostesses tend to frown upon guests
who extinguish cigarettes in Waterford bonbon
dishes! Also, if you must smoke, do be very, very
careful about where those ashes go. I know a
man who lets the ash grow very, very long,
and just as I think it's going to fall in his lap, he
gives it a flick in the very general direction of the
ashtray, leaving a clutter of ashes along the
chair, the carpet, and the end table. He's so
careless that Smoky the Bear needs to come up
with an indoor campaign just to save hostesses
from his sloppiness.

WHAT KILLS LOVE?

I'D PENALIZE A MAN TEN YARDS FOR BEING
INCONSIDERATE ABOUT THINGS AT HOME:

1   Tossing socks, shirts, and other articles of clothing
    on the floor; draping jackets and ties over chairs.
    One ultra well groomed man I know actually
    maintains that because he always drops his
    clothes in the *same place*, his wife doesn't mind
    picking them up! Ridiculous. Why should any
    woman with a high school or college degree
    spend her time picking up after an untidy
    husband? She has enough to do with meal
    planning, washing, ironing, vacuuming, child
    chauffeuring, kitchen cleaning, and on, and on,
    and on, to have to waste any time picking up her
    husband's clothes. If you habitually walk into
    your living room after work and fling your topcoat
    over the nearest armchair, shame on you! Retrain
    yourself to walk directly to the closet and hang
    it up! And if I ever caught you tossing soiled
    clothes under a bed, or on a closet floor, I'd up
    your penalty to thirty yards! A wife should never
    have to play hide-and-seek just to assemble her
    washing and dry-cleaning loads.

    As for putting your trousers in a top drawer—
    why? It makes a bedroom look sloppy; it
    certainly doesn't do the trousers any good, and it
    does, you'll admit, make it awfully hard for your
    wife or lover to get into that drawer to get to your
    billfold!

    On the whole messy-clothes syndrome, just
    reverse the male-female roles . . . mentally, that
    is. Picture your wife's stretch pants decorating
    the top dresser drawers. Picture yourself on your
    hands and knees, retrieving her panty hose from
    under the bed so you can wash them . . .
    picture yourself stumbling over a pile of soiled
    bras, pants, and slips tossed by the bedroom door.
    Enough said?

253

2   Leaving a messy bathroom. Even if you have a
    five-bath house, where did you get the idea you
    can spray toothpaste over the mirror, shaving
    cream over the basin, or something less polite to
    mention over the toilet seat or at the base of the
    toilet? (A FIFTY-YARD PENALTY FOR
    THIS LAST OFFENSE—can you really expect
    your wife to feel passionate toward a man who's
    as messy with his male equipment as a baby
    boy being potty trained for the first time? No
    woman deserves to clean up after you, and if you
    can't control your wonderful male organ, to
    keep it from spraying the bathroom facilities, I
    hold out very little hope for your prowess in the
    great game of sex!)
3   Littering. If you come home and leave your
    umbrella in the entrance hall, your briefcase on
    the kitchen table, your hat on the den counter, and
    the *Wall Street Journal* on the couch
    (incidentally, nothing dirties upholstery faster
    than newsprint), who are you to be yelling at the
    kids about the basketball left in the driveway? By
    the time you get home, your wife (or your maid)
    has probably put in a good four hours of
    housecleaning. If you litter and destroy her work,
    it's cause for murder! To put yourself in her
    place, suppose you had just spent four hours on a
    client's tax report and your ladylove entered the
    room and defaced each page with a big blob of
    India ink? Now do you understand how she
    feels when you litter her clean house?
4   Complaining about food. If you had chicken Kiev
    at noon, and find that she has fixed chicken
    Kiev for dinner, don't you dare say, *"Can't* you
    fix something else? I had that at lunch." I
    actually know men who throw small tantrums
    about duplication of meals. How easily these
    scenes could be avoided if every man in the
    world took the time to find what his wife or lover
    had planned for dinner. If you do find your meals
    duplicated, you can absolutely forget my

penalty if you make the following compliment to
your chef on the home front, "Darling, I had
chicken Kiev at noon—at Pierre's—and yours is so
much better that I think Pierre ought to go back
to the Cordon Bleu and learn to be as marvelous
a cook as you are."

5   Failing to call her from time to time. The above
situation could have been averted if you had
simply picked up the telephone after lunch,
discovered the Kiev duplication, and asked if the
trusty freezer couldn't offer up something
different for dinner. In the happiest marriages and
romances I know, the man checks in by phone
several times a day. She always knows if he's
running late, if he's in the mood for company, if
he's discouraged and needs an extra dose of
tender loving care when he gets home. In a really
super marriage I know, Jim picks up his office
phone several times a day just to call home and
say "I love you." It takes less than a minute of his
time but conveys hours and hours of loving and
caring to his wife! If you don't call home to let
her know you'll be an hour late for dinner, or
warn her that you're bringing a friend home, or let
her know about any change in your plans, you're
being a very thoughtless person in the game of
love. You certainly wouldn't treat a business
associate in this manner—and doesn't your lover
rate better treatment?

6   Playing little boy with your health. I know men
who disregard every order their doctor gives them.
It might be a matter of dieting, or getting
exercise, or quitting smoking. When a man pays
good money for medical advice and proceeds to
ignore the adivce, he's putting his wife in a
really nasty position. Suppose you've been told to
lose fifteen pounds. Your wife knows the doctor
told you to lose the weight. Result: She tries to
help you by fixing meals built around such items
as broiled fish, crisp salads, low-calorie green
vegetables. And what do you do? You snarl

because the fish isn't smothered in Hollandaise sauce, because the salad is topped with a mere teaspoon of vinegar and oil instead of a cup of Italian dressing, because the vegetable happens to be green beans rather than buttered corn on the cob. If you complain, what is your poor wife supposed to do? She's trying to help you stay healthy, and you repay her by griping about her efforts. You put her in the position of *having* to nag you. (On the other hand, if she doesn't nag you about watching your weight, stopping smoking, and all the other goals you're supposedly working for, you had better watch out. She's probably sick of the battle and plans to collect your insurance money much, much sooner than you ever thought.)

I'D PENALIZE A MAN FIFTEEN YARDS FOR COMMITTING THE FOLLOWING FAULTS IN PUBLIC:

1 Failing to make reservations before announcing, "We are going to Super Gourmet's for dinner." How embarrassing. Haven't you seen couples or whole parties of people kept waiting and waiting in restaurant lobbies because the host had neglected to make a firm reservation or show up on time to claim it? Remember, a woman dearly loves and admires competence in a man, and if you blunder by not prearranging an evening, you are automatically booted back to the Little League in her book.

2 Showing off in public. A man who is confident and sure of himself doesn't have to speak in a loud voice, order waiters around, send food back to the kitchen, return wine to the cellar, or do anything to call attention to his great degree of "sophistication." (If service or food or drink should prove to be less than acceptable, you should certainly deal with it, but the penalty holds if you do so in a loud voice.)

3 Letting a woman pay more than her share, or

being unprepared with ready cash when you take
her out. If you want to kill love before it even
starts to flower, try borrowing money from
the woman you're dating. Or try taking her out
to dinner and not having enough to pay for the
meal. If you're married (how did she ever get
involved with a guy like you?), continue these
bad habits to kill her love. Bounce checks. Insist
that she keep working, and squander her money
away. (If you both work, you ought to be seeing
some visible benefits from having two salaries.
If you can't work out a budget that lets you
save, hire an accountant to do it for you.)
Remember, whatever your situation, nothing,
NOTHING turns a woman off faster than a man
who's always short of cash, always sponging off
her, or always borrowing from her. When you
do this, you're practically screaming "You
Tarzan. Me Jane."

4   Making a big deal of paying the bill. One woman
I know said that the most awful thing her
ex-husband did, the thing that killed her love
faster than any other factor, was his insistence on
being Mr. Big Shot. "It embarrassed me so much
I dreaded going out with him," she said.
"Wherever we went, as long as there was
someone along besides me to appreciate him,
Matt would loudly summon the waiter and make
a big production of taking a wad of bills from his
wallet to pay the check. He overtipped terribly,
even when service was poor—just as long as
someone was around to notice him. He even had
to big-shot it by loudly sending drinks over to any
near celebrity in the room. He showed off so
much I finally told him I'd rather stay home. At
home, of course, it was a different situation. He
kept me on a strict budget, and our house
always needed painting. His car was always the
flashiest thing on the market, but there was never
enough money to let the kids have piano
lessons or go off to camp. The fact that he

257

lavished money on near strangers while virtually
ignoring his family hurt me so badly that I grew
to hate him."

5   Sitting on your hands when the bill comes
around, or failing to meet your entertainment
obligations. This is the reverse of Matt's
behavior, but it's just as much a killer of love.
Jerry, who has a hefty bank account and a lovely
portfolio of investments, is always tying his
shoelace or visiting the men's room when the bill
comes around. Carol, his wife, is sick of his
penny-pinching. She senses a distinct coolness on
the part of their closest friends, who have now
been stuck with dinner checks four consecutive
times. She'd rather have less money in the
bank and be able to face up to their obligations
as a couple. "It's bad when we go out," she told
me, "but it's even worse at home. We seem to
eat steak and have nice wine and graciously
served drinks before dinner when we go to other
couples' houses. When I entertain, Jerry buys
the cheapest liquor he can find and rations it
out as though it were the last in the world. I'm
a good cook, but I hardly think the people we
entertain go for tuna casserole on Saturday night.
Oh, it's not quite that bad, but almost. I've
reached the point where I can hardly hold my
head up around our friends. I know they talk
about Jerry's stinginess behind his back. He's a
good manager, but all that money in the bank
isn't doing us much good if I'm constantly
having to act as though we're paupers."

6   Reliving, and reliving, and reliving those dear old
days when you were the prankster-in-residence
of Alpha Sigma Sigma fraternity. When you
and the other ASS members get together, it is
boola-boola rah-rah-rah all the way. Your wives
want to scream, and I think they've got every
right to do so. We're all glad you spent such a
hilarious four years in those hallowed halls of
learning, but if all you can talk about is college,

258

you're really advertising the fact that you
haven't done one noteworthy thing since you
grabbed your sheepskin and departed. If you
must reminisce, at least spare outsiders the
experience and confine the sharing of those
memories to alumni gatherings.

7 Tearing down or embellishing someone's stories.
You're most likely to do this with your wife, but
some men I know apparently can't stand to let
anyone else have the floor for any period of time.
I vote for *never* interrupting anyone's story—
unless it's to defend a friend who is being
absolutely slandered by the story in progress. As
far as interrupting, does it really *matter* that you
interject, "No, it was *Tuesday* when it
happened, not Thursday," or "Alice, the article
you're talking about was in *Time* not *U.S.
News.*" You may be right as rain, but you'll never
win any friends with your constant corrections.
As for your wife, let her tell it her way in public.
If you think you *must* set the poor girl right,
wait until the party's over and tell her at home
(by which time the correction will assume its
proper status and will probably not be worth
mentioning).

8 Lying or exaggerating. When you catch a person
in even a small lie, how do you feel about him?
(or her?) Killing someone's trust in you is a very
effective way of killing love. It's so much easier
to tell the truth. You'll sleep with a free
conscience, and you'll never have to work your
way out of those "tangled webs" you create with
falsehoods. I think a lot of men, and women,
too, would have a much easier time of life if they
learned to say No. If you know darn well you
have no intention of attending Mr. and Mrs.
Bore's cocktail party, say No, and say it as soon
as you receive the invitation. You can say, being
very, very honest, that you "have other plans"
for that Saturday. Never tell a white lie and
conjure up a business trip or a specific place

you must be. Instead, be perfectly truthful with the "other plans" gambit. The Bores don't have to know that your "plans" at the moment consist of no more than the Saturday night movie on television. If the Bores are nervy enough to press you as to what your specific plans might be, you are under no obligation whatsoever to tell them. Just remember that your time is much too valuable to waste in elaborate deceptions.

As far as exaggerating, we all do it from time to time. But if you've got a really entertaining story to relate, it ought to be just as good, and far more believable, without exaggerating all the way through. Tell it like it is, and you'll have listeners who won't turn you off the next time you have something to say.

9 Lingering too long at the cocktail hour. Not a single woman I interviewed felt that her husband performed better in sex after drinking. If you carry your drinking to extremes, your performances in bed will drop, and drop, and drop. You may *think* you're great, but you couldn't be more wrong! And, aside from impairing your sexual abilities, you're going to reek like a distillery, which is a bad, bad turnoff from the very start! If you persist in drinking too much, it won't be just your sex life that's affected, you know, but your great mind, your performance at your work, and your general physical health.

10 Neglecting care of your teeth. Alcohol breath, coffee breath, cigar and cigarette breath, tobacco-stained teeth, unsightly fillings, the after-garlic and after-onion breath—what turnoffs to love! Your dentist can really be your best friend in helping your sex appeal. Spare no effort to make your breath as fresh as possible and your smile as handsome as can be.

11 Failing to observe personal cleanliness. So many women have told me their husband's feet *smell*. Now, could you possibly love a woman who had

beautiful hair, lovely eyes, a great figure, a
general aura of Chanel No. 5 around her
uppermost regions—and *smelly feet*. How
frightful!

12 For committing some of the more grievous sins
against etiquette. Real turnoffs for love are
failures in observing some of the simple rules of
etiquette which make our world just a little nicer.

I still see men who ought to know better doing some
pretty awful things, like using a toothpick at table or
when walking out of a restaurant (ugh). If a piece of
meat or other food should lodge between your teeth, you
do know where you can go to remove it, don't you? There
are also men who have tics and itches that they seem to
feel they must relieve in public. One very attractive man
I know blew his whole nice image every time he returned
from the men's room. His left hand was always fiddling
with his privy parts, as though rearranging them in his
trousers! As with annoying habits, his gestures were so
predictable that I soon ceased to think of him as "——,
that very talented editor," and began to think of him more
in terms of his bad habit. He became, and always will re-
main in my mind (although he's made the big time in
news agency reporting) "—— with the Crotch Problem."
Even now as I read his excellent news reports from Tan-
giers, from the Paris peace talks, and from developing
African nations, I picture him emerging from men's rooms,
left hand busily doing whatever it is that he does after
urinating!

Such sins against polite behavior tend to overshadow
many nice qualities about a person. You can become
"Gene with the rasping cigarette voice," or "Fred with the
throat-clearing habit," or "Charles who always has that
nasty cigar in his mouth," or "Eddie who's always interrupting."

As much as I hate to mention it, there is also the fellow

who, alas, always seems to be, to put it bluntly, passing
gas through his rectum. How terribly human! But how
humane and thoughtful to leave the room, or if caught
without warning time, utter a soft-spoken "Excuse me."
Oh, what a turnoff for love. Yet would you believe I get
constant complaints from wives about this? One woman
told me, "My husband does this constantly. He does it
loudly, and then, in a very shocked voice, reprimands our
innocent little dog for making the noise. What's more, I
think my husband thoroughly *enjoys* a you-know-what. It's
inexcusable. I'm tempted to name our next dog 'Fart.'" I
agree, and as they say after the patent-medicine ads, "if
the condition persists, see your doctor."

It came as a shock to me that each person I interviewed
could tick off dozens of turnoffs for love, almost as if he or
she had them ready and on the tip of the tongue for a
long, long time! They had very strong feelings about the
no-nos that ruin a relationship.

These same people, when asked to enumerate the nice
things that attracted them to another person, spoke much
more slowly, and in much more general terms . . . items
like . . . "I like a tall man, with a good personality . . .
someone who has a good sense of humor"—that sort of
thing.

It made me decide that being an attractive person, an
attractive lover, is a very complex goal, BUT to succeed, a
man must first eliminate the negatives! If women feel
such strong distaste toward men who show off, who drink
and smoke too much, who are careless and inconsiderate
toward others, a man must first eliminate every one of the
commonly observed faults in this chapter.

Turnoffs for love are very much like using bad grammar. I've certainly never heard a woman say, "Ooooooo,
that divine, divine Ralph . . . he uses the most beautiful
grammar I've ever heard." But let "divine Ralph" mouth
something like "She don't never want to set near the front
of the picture show," and he's blown his image to pieces
immediately.

Your use of good manners, of consideration, of careful grooming are rather like Ralph's grammar; your blunders will make an immediate bad impression and be an instant turnoff for love. Your smooth and proper knowledge of all the niceties might not make such an *immediate good* impression, but if you master all the right things to do, you'll be building your reputation as a good lover and good person for the rest of your life.

Chapter 15
How To Be A Perfect Husband —Although No One's Perfect

Every woman knows a perfect husband—the only trouble is that he's always married to someone else! In all probability, the man she thinks is "perfect" as a husband is someone whom she doesn't know too well but observes from afar as he goes through life doing what she thinks are all the "right" things. He still opens doors for his wife, and he's the first one on the block to cut the lawn, and he's off to church every Sunday, and he's a Cub Scout pack leader, and he's perfect—as far as she can see. If she could look into his private life with his wife, or knew his wife better, she might learn that Mr. Perfect was lousy in bed, was a bore when out among others, and tended to nag her constantly about her housekeeping!

No one's perfect. Some men have wonderful qualities that they just fell into, like a perfect profile, natural grace, and a large inheritance.

Other men have to work at their appearance, their financial status, their charm. Some of the men who have to work the hardest make the best husbands!

Since we've decided that no one is perfect, and you're breathing a bit easier, let's talk via tape with some of my nominees as "almost perfect" husbands.

These men are batting as close to 1,000 as they can and still be human.

In my search for a "perfect husband," I came across a man who won the title—as proclaimed by his lovely wife of sixteen years. Her total happiness with her mate is reflected in her shining good looks, her joy in everyday living, her obvious satisfaction with her two children (one eleven, one fourteen), and her friends.

Gene, her husband, is in his mid-thirties, has his own mortgage firm, nets about seven thousand dollars per month, keeps his splendid physique by regular exercise, and works harder at his love life than does any other candidate I've ever interviewed.

"If any guy is making love to his wife in his thirties the same way he did when they were in their twenties, he's not doing his part to keep love alive," says Gene.

Gene is ready and proud to announce to one and all that he has never been unfaithful to Julie in the entire sixteen years of their life together.

"Anyone interested in keeping love alive has to go beyond position one to do so," Gene maintains. "But I don't have to sleep with a lot of other women to learn these techniques. I can get all the education I need at X-rated movies and can find plenty of ideas to bring home to try.

"I might go home and mention something I've seen and get Julie's reaction to it. If she doesn't like it, I learned to let the topic of conversation drop for a week or two. This lets her off the hook. Then in a week or so I might bring up the idea again. Sometimes she will have changed her mind, and we can try the new thing out.

"When we were first married and I wanted to try oral sex, it took a long time to get her ready. We talked on and off for a couple of years. I took my time. But when she was ready to try it, it was a very pleasurable experience for her because both her mind and her body were attuned to the idea.

"I guess you might say I've made my wife my mistress, and I love every minute of loving her. I know I'll never have to say I started up with another woman because my wife wouldn't give me what I wanted."

Julie, in turn, has used the same gentle tactics with her husband. Julie, from a strongly religious background, became deeply involved with a local church group which Gene tried to avoid. But because they talked over each other's desires so thoroughly, Gene never begrudged her the time she spent with the church's activities.

"I think it was because Gene was so kind and understanding about my feelings toward church," Julie told me, "that I found myself beginning to bend a little. For instance, if a football game that Gene really wanted to see started early, I stayed home from church and watched the services on television before going with Gene to the game.

"He taught me to be tolerant of his needs. In turn, when I didn't raise a fuss about those few football Sundays each season, Gene respected me for it.

"Finally, he started coming to church with me because he wanted to share it with me. If either one of us had been petty or intolerant of the other's Sunday plans, we could have ended up sulking and mad, and apart a lot of our Sundays. Instead, he now goes to church with me almost every Sunday. When a game comes up that he wants to see, and it falls on Sunday, we do that. But we're doing it together because his tolerance prompted mine and things worked out."

This tolerance and giving carries over into every phase of their marriage.

"I discovered when we first started dating that Julie was a night person who prefers to sleep until ten in the morning," Gene told me. "After the children came, I thought, heck, I get up early every morning anyway so why should she have to get up too? I fix breakfast for the kids and me, and never give it a second thought. She has more than made it up to me by being a wonderful housekeeper, supervising great meals for company, and still being bright eyed when I come home in the evenings, and late into the evening."

267

Their togetherness will last a long, long time, I think. Their children appear to be as "perfect" as their marriage. Both are almost A students and are given the marvelous security of a happy mother and father. (And if the children's personalities are warped by having Daddy fix their breakfasts, they certainly hide it well from the world!)

Gene and Julie guard their privacy well. "We have a rule," says Gene, "that after the children go to their rooms at night, no child knocks at our door unless it's a dire emergency. We have our time with them, and after they go to bed, we have our time for ourselves. They don't seem to be hurt by this; they understand our need for privacy."

Gene is a manly, but gentle, type. I suspect the reason he deals so well with marriage is that he is perfectly secure within *himself*. His considerate attitude toward his wife seems to extend to his wife's friends and activities as well.

He commented on his wife's everyday life. "Anything Julie wants to do in the daytime is fine with me," he said, "and she seems to be in a whirlwind of activity with church, women's groups, and the children's activities. But I've always asked her to reserve the evenings for me. She has a couple of gabby friends, and if they call at night when I'm home, Julie asks them to call back in the morning because she's 'busy.' The same goes for nighttime activities. I don't see any reason for a woman to be doing something like playing bridge with the girls at night—unless her husband has a meeting he has to go to.

"Couples belong together, and we try to spend as much time together as we possibly can.

"We try to find some special times for lovemaking, like after the children have left for school in the morning, or occasionally at noon when I come home for lunch. Making love at a different time gives us pleasure."

Gene, a man successful in his marriage and in his business, makes it a point to share business triumphs and problems with Julie.

"I like to think that we both do better when Julie understands what I'm trying to do," Gene says.

"Each of us is better than the other in different ways. I'm better at financial matters and judgment in family spending. She's more sensitive and emotional and has a great artistic sense.

"She designed a new logo for my company, and I bring employee relations problems home and we talk about them because she seems to understand how each person in the corporation feels, and she helps me make the right decision.

"On the other hand, I help her by giving her an easy budget to live with. Once I told her how much of our income should be spent on clothing, housing, and so on, and she took to the budget immediately, and has never exceeded it."

Because Julie understands what Gene is doing, she's never in a bad mood when he calls and says dinner must be late because he's going to stay with clients until a contract is finalized.

Gene summed up our interview by noting some of the troubles experienced by his married friends. "Men so often make the mistake of assuming that their wives will accept their new ideas about lovemaking before they give the wives time to get used to the new ideas.

"Things like that take time. But I've seen husbands just shrug their shoulders, assume that the wife is less than passionate, and seem to think that the situation somehow justifies their starting an extramarital affair.

"It's sad, because usually everyone gets hurt in such a situation. If more husbands talked out their ideas on lovemaking, and gave their wives every chance to get used to new techniques, and lots of time to become accustomed to them, there wouldn't have to be any running around at all. They'd all be like me, with my great mistress who just happens to be my wife and the mother of my children as well!"

ANOTHER VIEWPOINT

One of my nominees for "perfect husband" described his marriage this way.

"My years with Sally I've looked at as I would building a great business or a great house. I think all men who are positive about their marriage do this. Let's take the house. It might start out as a new house, and everything about it is fresh and exciting for a few months," said Phil.

"Then something has to be fixed, minor repairs at first. Then several rooms have to be repainted. After a couple of years, maybe the entire outside of the house has to be redone if you want to keep the house in good condition.

"You have the choice of starting over with an entire new house, or you can stick with the one you've got, because you know it's basically sound, and keep putting work and effort into keeping it nice.

"If you start out with a pretty good house and never let down on its upkeep, in ten or twenty years you've got a house that's much, much better than some of the brand-new houses that haven't weathered a bit and increased in charm. (And you've saved a lot of financing and moving costs along the way.)

"I've looked at our marriage in just this way. I've never thought I could let things go or quit feeling as strongly for Sally as I did the days before we were married. Don't ever think that marriage isn't work, as wonderful a work as it may be."

Phil said that he thinks a man has to find out his wife's weaknesses and her fears, and then he can go to work to build her up. "I don't want a clinging vine, a wife who leans on me for every decision," he emphasized. "But a man has got to realize that women have special areas in which they can seem sort of insecure. If the husband can pinpoint these areas and try to build his wife up in these areas, she will be much happier—and as a result, he will be, too."

AN EXPERT'S VIEW

"The trouble with so many marriages," a psychiatric social worker told me, "is that both the man and the wife spend so much time pursuing some ideal that they've set for themselves in marriage. Often, it's an ideal that might have been great for their parents but is not at all right for them.

"As a result, they spend a lot of time feeling guilty that they aren't attaining the so-called ideal.

"And they waste so much energy feeling guilty they have little to use on moving forward, or attaining more reasonable goals."

Do you know how to fight with your wife?

Oh, I don't mean as in "Have you quit beating your wife?"

I mean that how you handle marital fighting, or spatting, or quarreling, or whatever you want to call it, is highly important to the success of your marriage.

Do you know couples who claim that they never fight? Either they're lying or their marriage is probably the biggest bore this side of Shadygrove Rest Home! It's normal for two people living together to disagree. If they don't, one partner is usually completely subservient, like an obedient but thwarted child, or the marriage blahs its way on through the years, with no downs—but no ups either!

How you handle disagreements makes a lot of difference.

TWO WAYS OF FIGHTING

Consider Sid Stupidfellow's handling of the following situation, versus the treatment given the situation by Joe Smartguy.

In a nutshell, both guys are confronted with very messy houses. They both suspect their wives of playing too much

271

bridge and paying too little attention to things at home.

Both guys at mad. Both want the situation corrected.

SID STUPIDFELLOW: "This house is the damnedest wreck in Teaneck! If your bridge game is more important to you than my breaking an ankle tripping over Tinkertoys, the heck with it."

JOE SMARTGUY: "Honey, Peter may be just four, but I think he's old enough to learn to get his toys out of the living room before I break my fool ankle on his Tinkertoys. The kids have got to understand that you need some time away from the house—and that they've got to cooperate so you can have it. I'll talk to Peter right now, and next Tuesday, before bridge, you remind him and the sitter to pitch in so things go better."

If you think the two different approaches aren't important, listen to what a psychologist told me.

"Using the 'I' approach in an argument can take a lot of the sting out of criticism. When a partner says, 'I think,' or 'I wish,' he's turning the controversy back toward himself. But when he constantly says 'you' or makes sweeping generalizations, the person being confronted feels threatened and accused."

Do you get the pitch?

Sid Stupidfellow's wife could correctly infer that Sid was accusing her of being the worst housekeeper in Teaneck. She might even feel that Sid had discussed her housekeeping deficiency with fifteen men at the office, with the corner druggist, and with every woman in her bridge group. In two sentences, Sid also managed to slam her activity of playing bridge and to imply that she was threatening his physical health.

Sid's wife could well get so mad that she'd toss it up to him that he hadn't had a promotion in three years, hadn't cut the grass in two weeks, and came home slightly tipsy from last Thursday's poker party . . . and the fight would be on!

Joe Smartguy, on the other hand, confronted with the same situation, used his head.

By starting out saying "I think," Joe turned the messy living room into a mere observation, *his* observation. He *did not* imply that the Tinkertoys on the floor were a crime of major import, one that the entire community was talking about. He managed to show his love for his wife by subtly demonstrating approval of her bridge game and showing that he *cared* that she had a day away from the children. Then he turned the situation even further away from mere criticism and suggested a concrete way to remedy the situation: that the child be encouraged to pick up his own toys so that Mother could have a little free time.

Come next bridge day, Joe's living room will probably be spotless. Sid's? Depending upon how long the argument went on, Sid might walk into a perfectly constructed Tinkertoy boobytrap—designed not by his small son, but by his still angry wife!

Another fault of Sid's in his "fighting" tactics is what my psychology consultant calls "gunnysacking" of grievances. Sid never got little grievances off his chest as they occurred. Instead, he'd wait until something (like tripping over his son's toy) provoked his temper. Then he'd launch on a thirty-minute recitation of his wife's recent shortcomings—and before it was over would usually mention "that no-good brother of yours and that mother of yours who never lets us have a minute's peace when she visits."

A composite of opinions from women I've interviewed has led me to believe that you'll earn a "perfect husband" title if you:

1 Talk over your problems as they happen and keep those vital communication lines open every minute.
2 Are frank with your wife about all your problems, including business. (If your self-esteem is low because of a business reverse, you'll be likely to take it out on her. If you explain what's bugging you, she'll be ever so much more understanding!)

3   Make little "contracts," either written or verbal.
    Spell out your agreement to do something that
    she wants and ask her to do the same for you.

4   Keep your criticisms ever so gentle. Example:
    "Honey, ninety-nine people out of one hundred
    would never notice because you always look so
    nice—but I loved your hair when you wore it in a
    less jazzy style."

5   "Reinforce" her ego rather than tear it down. Try
    to find things you love about her and TELL
    her about them.

6   Let her know when something she does displeases
    you but always in private, never in the company
    of others.

7   Admire the appearance of other women—ONLY
    if you have complimented her in the same
    manner.

8   Build up her strong points and emphasize her
    UNIQUENESS as a person and how much that
    uniqueness means to you. (In this stereotyped
    age, we *all* need to feel that we're different,
    that we have something to offer that no other
    person has.)

9   Compliment her in front of others, as well as in
    private. This makes her feel you're proud to
    share your good feelings about her.

10 Share responsibilities about decision making,
    money spending, child rearing.

11 Show how much you care when she's ill. A
    woman's nasty cold is half-cured when a husband
    shows his real concern. Just as you want large
    doses of TLC at some special times, so does she
    (especially when she hands out such large
    amounts to you and the children).

12 Make as much of an effort when you entertain
    "her" friends as she does when you entertain
    "your" friends. Every marriage turns up sets of
    "his" and "her" friends to be entertained. If you
    sulk or show you don't care when the evening is
    primarily for "her" friends, you're putting
    yourself in a terrible light!

13 Try to show her that she's more important than your business schedule. Longtime Speaker of the House John McCormack and his wife, in the many decades he served Congress, never once missed having dinner together. That's what I call love!

14 Remember that being "manly" and being "sensitive" are not two incompatible qualities! Real men, those who are perfectly secure in their masculinity, are never afraid to show deep feelings . . . yes, even to cry. I would hope you would bring your son up with this understanding, too. He'll be a better man if you do!

15 Put yourself in her shoes. (In an "encounter group" you would assume and act out her role to help you understand her feelings.) Of course your job is demanding. But if she's running a home with small children, the demands on her time may go on for eighteen hours a day—with no real lunch or coffee breaks, since those dear, demanding children are right there with her! The nicest husband I know, although he travels almost constantly, never fails to take his wife *out* to dinner his first night home. That first night home is a sacred thing with them—their time to be together.

16 Steal her away from the kitchen and the kids for one-to-three-day trips. Do this as often as possible, and make a pact not to discuss those wonderful children during your "honeymoon." Doing this insures that you continue to grow as two people together—the way you started out. It insures that your relationship is still blossoming as it did before the stork arrived with those precious bundles.

17 Learn that senility and middle age are more psychological than physiological. Keep growing, keep moving, keep developing new interests. Don't let yourself become a bore, and your stock as a husband will never diminish!

18 Respect her right to continue her growth as an

individual, too. Don't belittle her taking a
modern dance course at the "Y," or enrolling in
first-year piano lessons with your six-year-old
daughter, or taking a course in decoupage at the
hobby shop. Happy wives make for happy
marriages, and few women can stay happy with
nothing but a new can of acrylic floor wax to
occupy their minds!

19 Understand that women do have their ups and
downs. She most likely has a few days a month
when she's likely to be rather depressed—or
downright nasty. She doesn't like herself at these
times, either. Respect her right to be moody, and
try to be a little more patient when moodiness
does occur.

20 Call her from your office. Husbands who rated
tops in my poll called home two or three times
a day. The calls were brief—to share a funny
story, to ask if she needs anything, or simply to
say "I love you." Most women hesitate to call
their husbands at work, and this is proper I think.
But when you have time, a call tells her you
care. You'll *always* call if you find you'll be late
getting home. *Always!* This makes you much
more of a man!

21 Keep her informed. If you're to meet clients for
drinks before dinner, tell her! If you might be
held up in a conference, warn her! So many men
miss this basic rule of good manners, and the
ones who do are the same ones I find griping
because their wife "resents" their wanting time to
themselves.

22 Tell the truth. If you lie about small things, she
will never trust you about anything! If you had
lunch with a good-looking redhead who
handles buying for your television account, TELL
YOUR WIFE. Make your relationship so open
that neither of you has cause to doubt each other's
total integrity.

23 Pay as close attention to your grooming as you
did on your first date. Aftershave lotion and men's

cologne are still sexy aften ten years of marriage, you know. Brushing your teeth before lovemaking is a basic courtesy. So is shaving—if your stubble is such that you might scratch!

24   Work with your day-to-day communication. Write little notes. Write long letters, if you're in in the mood. Get things down on paper that you've always meant to tell her. Give her concrete evidence that you CARE!

25   Give her a hand when she's rushing to go out with you for an evening. Help feed the kids. Actor James Garner told me he'd become quite adept at fixing his daughter's hair and getting children ready for bed. Why can't you do the same? If you want your wife to look like Jackie You-Know-Who when you take her out, how can you sit there reading the evening paper while she feeds and bathes children, takes out the garbage, gets the kids into their pajamas, does the dishes, and picks up the sitter? You deserve to have a wife who looks like Gravel Gertie and smells of Gerber's spinach rather than Arpege if you treat her like this!

26   Fill her in on your financial situation. Explain your budget to her so she can help you make it work.

27   Either teach her how to handle her affairs should (heaven forbid) something happen to you or place your estate (no matter how small) in the hands of a bank trust officer who will take care of her if she's not equipped to cope.

28   Insure yourself. The most healthy of men can get beheaded by an errant Frisbee and die a premature and unplanned death. You've got to have your family provided for!

29   Take such good care of your health that some newspaper writer will be interviewing you on your health secrets in the year 2045. Your wife *wants you to be around*, you know, or she wouldn't have married you in the first place.

30   LEARN THAT LOVE MEANS NEVER

BEING AFRAID TO SAY "I'M SORRY."
Contrary to current nonthinking on the subject,
I say that people in love never lose face by
saying "I'm sorry." Furthermore, you don't have
to say "I'm sorry that I ———" and risk recreating
the fight. Simply taking her hand, or giving her
a hug or a gentle kiss, and saying a sincere
"I'm sorry" can heal little hurts before they
become big wounds.

You can build a happier, stronger marriage if you will just add one of my suggestions to your life each month. Your giving and thoughtfulness will be repaid tenfold, and your marriage will be revitalized.

(Don't tell your wife about your plan. Just get to work and carry out my suggestions, adding one new one each month. Keep using them, and watch these ideas become the mortar that cements your marriage with new strength and new rewards. This is our secret, remember! And remember, also, that after you practice the suggestions for a while, they'll become as much a part of you as your love equipment.)

After you see the changes in your *own* marriage, share the suggestions with a friend and hope that *he* shares them with *another* friend. Maybe we can build a pyramid club of thoughtful husbands and close every divorce court in the country in the next five years!

All the ideas you've read about your physical condition, your grooming, and your techniques in bed will make you a better lover longer.

Once you've mastered the thirty suggestions on how to be a better husband, you'll make her want you Every Day, in Every Way.

Chapter 16
What Is This Sex Thing All About?

We're almost ready to leave the playing field now, and I hope that you've learned expertise in a number of new plays and are determined to follow a program of self-improvement that will make you LOOK like a much better player.

The game I really want you to star in is that of SEXUAL LOVE, so let's think for a minute what this thing called sex is all about.

I know that in the locker room you've heard a lot of other players bragging about their past games or the game that they're planning with a new partner tonight. The sad thing to me is that they're playing a very different game from the one I want you to star in. These players are participating in LOVELESS SEX.

After they've played LOVELESS SEX and you play SEXUAL LOVE, two sets of experience are going to result in two very different trophy rooms.

YOUR TROPHY ROOM
FOR SEXUAL LOVE

Your walls will be covered with blue ribbons and with pictures of the very special woman or women with whom you've experienced love. The ribbons are inscribed "FOR UPHOLDING THE BEAUTY OF THE GAME OF SEXUAL LOVE" and "FOR A SPORTSMANLIKE CONTRIBUTION TO THE BETTERING OF SEXUAL LOVE."

Your photographs are inscribed "I'll always remember you fondly," and "It was wonderful playing with you," and "With thanks for our beautiful lovemaking . . . never to be forgotten." The picture in the place of honor is signed "To the man with whom I want to spend the rest of my life," and its date is the most recent.

Your trophies are inscribed "For Achieving High Self-Esteem Through Expressing Sexual Love," and "It Matters Not Whether You Win or Lose, But How You Play the Game," and "Lifetime Achievement Award: Building and Maintaining Sexual Love With Meaningful Relationship to Special Partner."

HIS TROPHY ROOM
FOR LOVELESS SEX

His collection of mementos and "awards" after ten years of playing at the sport of LOVELESS SEX is very different. On his coffee table is a large scrapbook *filled* with receipted bills.

Lots of the bills are from motels. Each one is carefully annotated. "Laura—a loser, but an easy lay." "Susan—she turned me on for two weeks." "Beverly—a bitch, but red-hot in bed." "Janet—a challenge to get to, but no big deal once I did." "Sondra—too sweet to get a bang out of sex."

Two bills are each in excess of $500. They're for abortions. Several more are to a local jeweler. They're annotated "Luring-into-bed present for Lucille," "Buying-off present for Patti," and "Gift for hooking Carrie."

Two accounting sheets total a staggering $65,000. One lists, among other things, "Lawyer's fees—divorce from Betsy . . . property settlement . . . lump sum house payment for Betsy," and totals $41,900. The other accounting sheet, showing a total of $23,100, is divided among lawyer's fees, temporary support, and temporary alimony payments for "Jill and the kids."

A bunch of the bills are for cases of liquor (some of these are marked "for getting Laurel smashed and in the mood," and some are marked "for my nerves"). There are a lot of drug bills, too, mostly for tranquilizers and sleeping aids.

Dated quite recently are a series of receipted bills paid to Dr. L. Maurice Waldbaum, psychiatric counseling. At the bottom of the stack is a receipted bill paid to the Manly Arts Bookstore for a volume entitled *Screw Your Way to Happiness Tonight*.

On the walls of his trophy room, our star in Loveless Sex has a collection of letters. The salutations are rather unorthodox—things like "Dear Cad" and "You Damn Bastard."

One framed telegram addressed to our Loveless Sex participant says simply:

"Drop Dead—Karen"

YOU'RE THE PLAYER,
BUT IN WHICH GAME?

Both games—SEXUAL LOVE and LOVELESS SEX—are played with the same equipment, and they might even look pretty much alike once the game comes close to its natural climax.

But LOVELESS SEX, bluntly, is just a glorified form of masturbation, and you'll get just about as much uplift from masturbation as you will from loveless sex once you've completed the inning.

If you decide that you want to play the game that gives you a lifetime of gratifying, heartwarming experiences, seek out the real depth of love. Seek out the love that poets compare with breathtaking sunsets. Seek out the love that poets compare with the beauty of the moon's rays playing across a silvery lake. Seek out the love that the poets compare with the glory of a rainbow. You should know in which game you'll find this kind of joy. You will find it in the game of SEXUAL LOVE by learning and following all the rules, and by making the game as beautiful for her as it is for you.

SEXUAL LOVE finds you filled with that exhilarating sense of having become one with her, of having partaken in a true communion of your mind and hers, of your body and hers. In SEXUAL LOVE, you think not only of "me," not only of "her," but of "us." The movements and thrusts and ups and downs and ins and outs of both games might appear very similar, but how you *perceive* the game will either make it a beautiful one (SEXUAL LOVE) or an unsatisfying and degrading one (LOVELESS SEX).

I hope you seek the fulfillment that is SEXUAL LOVE and seek it with all your being. I want you to find the partner with whom you can have the wonderful experiences and feelings and depths that only SEXUAL LOVE can give you. You will understand why I feel this way if you take a piece of paper, head it "WHAT I WOULD LIKE TO GET FROM SEX," and attempt to list your goals in this uniquely physical-spiritual act. Then read below, and decide which way you'd like to think about sex, based on your self-evaluation.

Ideally, every one of your feelings and thoughts about sex will be more closely mirrored in the left-hand column than in the right. If you find yourself aligned more with the speaker in the right-hand column, you're still groping. You may be playing *at* sex every single night, but you haven't even begun to grasp the meaning of what this sex thing is all about.

On the other hand, if you identify strongly with the statements in the left-hand column, your perception of and sensitivity tell me that you have a full understanding of what this sex thing is all about.

282

SEXUAL LOVE FINDS YOU THINKING	LOVELESS SEX LEAVES YOU THINKING
"I melted into her, and we became one."	"I screwed her good."
"The next morning we just held each other close and felt this fantastic sense of achievement."	"I couldn't wait to get up, get my shower, and get the hell back to my own apartment."
"We want to be together every hour of the day."	"She's great loving, but I couldn't stand to live with her as a steady diet."
"She brings out tenderness I never felt before."	"Boy, she turns me on physically—but that's all."
"For the first time, I find myself saying 'we' and planning that way for the future."	"I'm not too keen on the way she wants to sort of pin me down about the future."
"The very thought of her gets me very much in the mood for love."	"A movie I see or a book I read gets me fired up, and then I think of her (or of Nancy, or Jill, or Margot, or Anita, or Joan)."
"When I'm with her, I feel ten feet high, and I want to show the world she's mine."	"She does a lot for me physically, but she's not, ah, the sort of broad you'd take home to Mother."
"Ever since I've known her, I feel I can conquer almost every situation that comes along."	"She was a good conquest; I played my cards right and came out on top."
"I don't love her because I need her . . . I need her because I love her."	"I need her when I need to make love."

283

SEXUAL LOVE FINDS YOU THINKING	*LOVELESS SEX LEAVES YOU THINKING*
"I love her body so much and appreciate its beauty more and more every day. Touching her breasts is a pleasure and joy all to itself . . . not just a prelude to intercourse."	"She's stacked, and her body is a great plaything. It turns me on."
"It's a privilege to give her oral sex and show her that I love every part of her body."	"I have to give her oral sex, because if I don't, I know she won't give it to me."
"The thought of someday having a child with her is beautiful . . . it would be something that was a part of each of us, and an image of our love."	"Boy, she sure better stay on the pill. I hope I can trust her not to make a mistake. A kid of hers is the last thing in the world I need."
"Just as I love every beautiful portion of her body, I want to give her every portion of mine, for all time to come."	"I'm mad for her—until next weekend when Doreen blows back into town."
"Entering into the mystery of her is the most fantastic thing that could ever happen to me."	"I can't wait to get into her panties."
"If she is hesitant about lovemaking, I make it plain that I'm willing to wait until she's ready. She has moods, and I respect them because they're part of her personality. It's easy for me to say I'll just hold her tonight and we can make love some other night."	"If she didn't get with it every time I needed her, I should sure as heck get another woman—and she knows it."

284

SEXUAL LOVE FINDS YOU THINKING	*LOVELESS SEX LEAVES YOU THINKING*
"Our sex life takes on many moods. Sometimes we are aggressive; often we're highly passionate; sometimes we're overly sexy; sometimes playful, sometimes laughing, sometimes reverent."	"She always gives me a release in bed."
"The first afternoon I found her, I knew my seeking weekends were over forever; we had found each other, and I knew it was for life."	"We started having intercourse one boring Sunday, and she helped take my mind off the football game I couldn't get tickets to. Next weekend, I don't know who the girl will be."
"I'm like a little kid, discovering something for the very first time. I can't get over the marvel of the newness of her each time we're together."	"You've seen one woman— you've seen them all."
"We spent one weekend in a summer cabin with some friends—we wanted sex very much, but it wasn't private enough so we walked out on the pier and just held each other close, and listened to the waves, and thought about our being together the next night, when we could go to all the depths of our love for one another."	"What a bummer of a weekend. We were trapped and she said we couldn't sleep together, and she suggested just going out to get away from the group and listen to the sea. I wasted thirty-six hours and got nothing but a sunburn."

You can choose the game you want to play, and you can choose the happiness and sense of complete contentment and high esteem that playing the right game can give you. You can fill your trophy room with well-earned tributes to your loving—or with a sad collection of your setbacks as you unhappily move from one bed to another in vain quest of lasting love . . . but never giving of yourself enough to build a relationship that gives you a reason to live each day in love. You can chalk up any number of temporary thrills and be left empty handed at the end of your allotted time for living and loving.

If you seek this sad alternative of LOVELESS SEX, you can "make love" without the least understanding of what love really is. You can have loveless sex and still achieve orgasm—but it won't ever be the wide-screen extra-special, super-fantastic chromocolor orgasm that SEXUAL LOVE can give you, and give you, and give you— with the same cherished partner.

If you build your game right, and play it with pride, your trophy room will begin to fill up with all the beautiful and happy souvenirs that your love can bring to your life.

Many athletes have trophy rooms like this, and in their old age, they can sit back and recall the glories of games once played.

I hope your fine playing will build up a beautiful trophy room, too. But even more beautiful is the fact that SEXUAL LOVE is a game that you'll never, never have to give up. If you play the game right and keep your physical fitness up to professional standards, you can enjoy the game for the rest of your life.

That's the reason SEXUAL LOVE is the most wonderful game of all. What a sport! What a glorious situation! You have an entire lifetime ahead of you. Now can you see why it is so very important that you become a BETTER LOVER . . . LONGER?

With all good wishes for your success,
Debbie Drake